AF227965

LIGHT

EXPLAINED

HISTORY
SCIENCE
CONCLUSIOON

Dag Landvik

Author's Tranquility Press
Marietta, Georgia

Copyright © 2021 by Dag Landvik.

All rights reserved. No part of this publication may be reproduced, distributed or transmitted in any form or by any means, including photocopying, recording, or other electronic or mechanical methods, without the prior written permission of the publisher, except in the case of brief quotations embodied in critical reviews and certain other noncommercial uses permitted by copyright law. For permission requests, write to the publisher, addressed "Attention: Permissions Coordinator," at the address below.

Dag Landvik /Author's Tranquility Press
2706 Station Club Drive SW
Marietta, GA 30060
www.authorstranquilitypress.com

Ordering Information:
Quantity sales. Special discounts are available on quantity purchases by corporations, associations, and others. For details, contact the "Special Sales Department" at the address above.

Light Explained: History, Science, Conclusion / Dag Landvik
Hardcover: ISBN 978-1-957208-52-7
Paperback: ISBN 978-1-957208-51-0
EBook: ISBN 978-1-957208-53-4

Contents

FOREWORD by Ingrid Fredriksson

In the same way as **Isaac Newton** (1642–1727), probably one of the most important scientists in the history of mankind, saw the apple fall and discovered gravity, Dag Landvik saw light reflect on the surface of the water and discovered that light is always perceived to make its way directly to the individual onlooker. He concluded that the light creating that glittery path on the water is in fact not created by the sun, but is just experienced or perceived as light inside the eyes and brain of a living creature. Pythagoras had probably made the same observation and had most likely come to the same conclusion, but Pythagoras 'conclusion was to be called the "emission theory", which in fact is somewhat misleading. Newton meant that light consisted of particles; different particles for different colours, even if he believed that these particles could give rise to waves in the "ether".

Huygens (1629–1695) advocated that light is a wave. EM waves are, as far as he was concerned, the explanation of light. Huygens' theory was not accepted or easily taken to heart, because Newton 's particle theory gained greater confidence due to his fame. Huygens' wave theory was however confirmed by

Thomas Young (1773–1829), who at the age of 13 mastered no less than 13 languages. The wave theory did not win until 1802, when Young showed that light passing between two thin slots created a so-called interference pattern, characteristic of waves.

With **Maxwell's** (1831–1879) equations, it was found that EM waves had exactly the same speed as light. Consequently, light was an electromagnetic wave with different frequencies for different colours.

Then there was **Max Planck** (1858–1947) with his quanta, later added to by Einstein with a stream of particles and photons each having the energy $E=h\nu$.

Many of the international scientists were, just like **Bertrand Russell** (1872–1970), also philosophically educated. Bertrand Russell explained his principle of light, the principle that can be considered valid for all other sensory experiences of humans. Since Bertrand Russell's explanation of the principle of the senses (1912) was based on EM waves and was put forward a few years after Einstein's already accepted particle theory of both light and the atom, Russell's explanation of light as EM waves was no longerconsidered relevant.

Louis de Broglie (1892–1987) meant that atoms are continuously receiving and emitting EM waves, which he later demonstrated with an experiment in 1927 and for that he was awarded the Nobel Prize in Physics in 1929.

Schrödinger (1887–1961), primarily known for the famous story of the cat, dead or alive, developed de Broglie's idea of the wavy nature of matter with his wave mechanics and wave equation.

So, what do scientists of today have to say about this? Light not existing in itself clearly does not tally, as there are currently advanced experiments being carried out on the quantum mechanical construction of light, where both generation and detection has been proven without any involvement of the living creatures. But according to modern-day science, the results of these quantum-mechanical experiments are difficult to understand as our brain works in a classic fashion. Interesting scientific research shows us other possibilities; for example, that birds use eye protein to "see" Earth's magnetic field, or, at the MIT university research is now being carried out to investigate whether facial movements can ask a question and the response can be given by vibration inside the ear.

This is where we are today when **Dag Landvik** makes his discovery. He simply and ingeniously sees the light reflecting on the surface of the water and discovers that light is always perceived to make its way directly towards the individual onlooker. He then concludes that the light creating that glittery path on the water is in fact not created by the sun but is just experienced or perceived as light inside the eyes and brain of a living creature. No matter how many observers are involved, they all experience exactly the same thing – the light reflecting and the glitter path heading exclusively towards each one of them. It is clear, that after Dag Landvik's discovery, the real world is no longer consistent with the map!

Ingrid Fredriksson

INTRODUCTION by the Author

It started one autumn evening in 2014. On the way home, an unusually grand full moon rose high above the forest and stood out powerfully against a dark sky. It was a magnificent image. I stopped to be able to take it all in and look at the details, then considered the fact that the moon was so brightly illuminated by the sun,which was actually on the other side of planet Earth. Curiously the rays of the sun were not visible from the side. Instead, the sky was completely dark, apart from the distant stars.

Surprised by the sunlight's lateral invisibility on the way to the moon and because my home is located close to water, a few nights later I observed that the light reflecting from a lightning lamp on the other side of the bay created a glitter path on the water, which headed straight towards me. If I moved sideways, the glitter path followed me and carried on being directed straight at me. This was quite interesting, so I decided to investigate the reason for thisstrange behaviour of the light reflection. Thankfully I did not ask anyone that had graduated from university – myself, I quit school prior to learning the physical explanation for this behaviour of the "night sun", i.e. Aristotle's conclusion that the sun's "light ray"is only visible as light in a straightforward direction, not from the side.

The quest soon led me to Pythagoras, who lived during the 6th century BCE and was the first and greatest philosopher of all time. Apparently, Pythagoras had also observed and addressed the same problem in order to understand the sunlight reflecting on the surface of the water. Pythagoras reached the conclusion that man's natural understanding of light is incorrect and that sunlight is not real physical light, but is only perceived as light by man´seyes and brain. Even if nothing has been left in writing by Pythagoras himself, enough documentation has been written down by others after him and which contains information about Pythagoras´ light theory, which also was the answer to my own problem of understanding the sunlight´s reflection on the sea from the setting sun.

After 2.8 million years of human thought and consciousness, this book offers insight to the nature of light, namely it being inseparable from the viewer. Consequently, life and light stand in agreement to each other – no life, no light.

Dag Landvik

1. INTRODUCTION TO LIGHT

1.1.   The International Year of Light 2015

The United Nations (UN) declared 2015 "The International Year of Light" [1]. By proclaiming the "Year of Light" the intention was to increase awareness and pay tribute to the concept of light, a natural phenomenon that not only is nature's main energy source, but also enables living creatures to see. Light is therefore crucial to mankind and all other forms of life but remains scientifically unexplained. Man still does not know what light really is. Ironically, to be 'enlightened' even has a symbolic meaning in terms of knowledge. These and many other important reasons make the secret of light the most urgent to answer out of all scientific queries.

1.2.   Thoughts on light

"All these fifty years of conscious brooding have brought me no nearer to the answer to the question, 'What are light quanta?' Nowadays every rogue thinks he knows what it is, but he is mistaken." (Albert Einstein, 1951) [2]

"As the sun sinks toward sunset, the incandescent-white reflection acquires gold and copper tones. And wherever Mr. Palomar moves, he remains the vertex of that sharp, gilded triangle; the sword follows him, pointing him out like the hand of a watch whose pivot is the sun." (Italo Calvino, 1985) [3]

"He enters the sea, moves away from the shore; and the sun's reflection becomes a shining sword in the water stretching from the shore to him." (Italo Calvino, 1985) [4]

"Light is great, light is mysterious, light is useful. After all, they say everything started with light. When any revolution has happened in the history of science, light was always there. So it deserves a little attention." (Fermilab, 1999) [5]

"Light is at once both obvious and mysterious. We are bathed in yellow warmth every day and stave off the darkness with incandescent and fluorescent light. But what exactly is light? We catch glimpses of its nature when a sunbeam angles through a dust-filled room, when a rainbow appears after a rain or when a drinking straw in a glass of water looks disjointed. These glimpses, however, only lead to more questions. Does light travel as a wave, a ray or a stream of particles? Is it a single color or many colors mixed together? Does it have a frequency like sound? And what are some of the common properties of light, such as absorption, reflection, refraction and diffraction?" (Harris & Freudenrich, PhD, 2000) [6]

"The many faces of so-called glitter path are presented to draw attention to an optical phenomenon which is not only interesting from a physical point of view but is also important in other contexts. It is shown as well that this phenomenon may occur in many situations totally different from the wet ambiance." (Prof. H. Joachim Schlichting, 2004) [7]

1.3. The unexpected explanation of light

The sun's reflection on water heads exclusively towards the eyes of each individual onlooker. People's spontaneous thought is that the light reflecting on water is created by the rising or setting sun and that this illuminates the whole sea. But as the glitter path of the morning or evening sun is always perceived to be heading individually towards each onlooker, the conclusion must be that the path of light on water is not created by the sun itself. The only possible explanation is that electromagnetic (EM) waves from a light source are not real physical light but merely perceived as light in the eyes and brain of a living creature. At the time when this book is being written (2015–2018), EM waves are still believed to carry the light as real light [8].

1.3.1. Light does not exist as light. Light is

merely a perceptionby the eyes and brain of a living creature, which occurs whenEM energy waves [9] from intensely hot atoms [10] hit the eyes, either directly from a light source or by reflection, and enables the creature to see its surroundings. The EM waves of light are therefore only energy propagation from a light source. The more energy the light source has, the further its EM waves will reach. When these hit the eyes of a living creature, the receiver will experience a three-dimensional field of vision following the energy waves in return to the light source, or its reflection points. What is called "the speed of light" is actually not a measure of any speed of light, as there is no such thing as real light, but a measure of the initial phase velocity of the EM waves from alight source. When the EM waves have reached their maximum energy propagation, man will no longer perceive or experienceany light beyond that point.

1.4. By understanding the nature of light, new insights are gained into the sensory functions of a human being

1.4.1. Man is, like all other forms of life, integrated with the physical systems of planet Earth through his senses. The secret of light is indeed a complex issue but has essentially a logical explanation. To take on this challenge has had its advantages, not the least including a practical approach without any preconceived ideas and being spurred on with the encouragement of Einstein: "The whole of science is nothing more than a refinement of everyday thinking."

2.HISTORIC THEORIES OF LIGHT

2.1. From sun god to science

2.1.1. For a long time the sun was deemed a god [11] **by humans.** Helios, the sun god, was thought to ride across the sky during the day and spread his light as he went along. Sunday, the day of the sun god, is the traditional day of rest and is consecrated to the sun in most cultures around the world. Duringthe 6th century BCE, a new movement started in the western world with Thales and Pythagoras; the organised, logical thought process of man with an aim to understand how nature, mankind and everything works.

Philosophy became a knowledge movement in anorganised fashion, where highly educated, intelligent philosophers took on the task of trying to understand the world. Understanding light was the first challenge. Light not only makes vision possible, but also the energy of light is a prerequisite for vegetation, food, life anda suitable climate for humans and animals to live on planet Earth.

2.1.2. Thales [12] (640 – 546 BCE) is considered the father of science in the Western World. Thales was a mathematician and engineer trained in Babylon and Egypt, with special expertise in geometry that he then brought to Greece around the year 575 BCE. Thales was highly regarded and respected and belonged to "The Seven Sages" [13] (the seven wise men). This knowledge-based organization became a template for philosophers and was well-known for its wisdom through brief proverbs available to read at the Oracle of Delphi; e.g., "Know yourself" and "Everything in moderation". It was however mostly women, of which the Delphi Oracle is the most famous, who conveyed these words of wisdom. "Wise men" have historically existed in most cultures. Thales and his right-hand man, Anaximander, were private tutors to Pythagoras [14] and both of them taught Pythagoras mathematics, geometry and cosmology when he was 18 to 20 years old. Thales was also the one who inspired and introduced Pythagoras to his long stay in Egypt and Babylon, which were leading knowledge centres in the western world of that era.

2.2. Science arose from the art of philosophy

2.2.1. Pythagoras (570–495 BCE) was the first Greek philosopher. He grew up on the island of Samos in Greece, where today there is a statue of him in the harbour, pointing to the sky asa tribute to his explanation of light. After being educated by Thales,Pythagoras spent approximately 30 years in Egypt and Babylon, which at that time were the most advanced civilisations in the westernworld. Pythagoras' light theory [15], which today is generally consideredabsurd – that the light is created in the eye and the brain – is actually, contrary to the current understanding, correct in principle. The theory does not, however, state that light is emitted from the eyes, which today is incorrectly and ironically given as an example of how primitive the first philosophers must have been. The theory does, in fact, describe what is perfectly correct – which is that the experience of light is created by EM energy waves emitted from highly heated atoms and that when these waves reach the human eye, the brain perceives this as light. So humans and all other living creatures with eyes and brain are the ones creating the light, or rather the experience of light. This is done by EM energy waves, which are not illuminated as such, but when they are emitted from the sun or other light sources and enter the eye, the brain will perceive light.

2.2.2. Pythagoras' philosophy was the starting point for western science. He is not only considered to have laid the foundation for mathematical science, but also for human development as fellow humans in a religious and social sense. Pythagoras' lifework was just as much about spiritual and social issues and behaviours as it was about mathematics and science and towards the end of his life it was conducted in a permanent academyin Croton, located in what today would be southern Italy. It is not known how Pythagoras' insight into the light question arose, but it islikely that the interest began in his younger years as a private pupil being taught by Thales, the father of philosophy. Pythagoras' education flourished further during his approximately 30-year stay in Egypt and Babylon. It is also feasible to think that his insightsand theory of light came from these cultures. It is, however, more likely that Pythagoras himself created his own understanding, as his theory of light has no known similarity in history. The problem with Pythagoras' light theory is that nothing has been recorded by himself. His main contribution to science is considered to havebeen mathematics. The insight into the mystery of light, however, was the most important and unique contribution. His explanationof light, i.e. that it is created inside the brain as an experience of the living creature itself, could not be proven scientifically without an appropriate theory to back it up. This theory did not become logical until much later, when light could be explained as EM waves.

2.2.3. The scientific way of thinking started with the attempt at explaining light. Since the sun, with its combined light and energy function, can be seen as fundamental to life itself, it was only natural that the first theoretical science, the art of philosophy, wanted to try to explain light. From about 600 BCE,the original light theory started to move away from a sun godworshipped by man, in an attempt to explain light on a logicallevel. The debate on whether the light occurs inside the human brain or externally in the air then continued amongst philosophers for nearly 1,000 years and only ended temporarilywhen religious groups took control of science around 500 AD. During the first era of philosophy, from Pythagoras (6th century BCE), Empedocles[16] (5th century BCE) and Plato[17] (4th century BCE), it was considered that the perception of light was created by man himself inside his eyes and brain, as per Pythagoras' theory. After this, during the 4th century BCE, Aristotle[18] proclaimed that light appear as light from external sources like the sun orsome other light source, just as we experience. It is also this "just-as-it-is-perceived" theory that has remained valid to this day. That is why man still does not know what light really is. Pythagoras had, however, been extremely close to the solution when he realised that light cannot be an occurrence in its own right but must be created as an experience or perception inside the human eye and brain. At that time it was not possible to fullyunderstand the concept of light as the philosophers of that era lacked the necessary knowledge of physics.

2.2.4. During the 4th century BCE, Plato and Aristotle proposed a "geocentric" world picture, which put planet Earth at the centre of the universe. This theory was well received, as it was an attractive idea for mankind. Aristarchus [19], a Greek philosopher, astronomer and mathematician, then proposed in the 3rd century BCE that Earth is actually orbiting the sun. Aristarchus' "heliocentric" theory, with the sun at the centre, was initially not supported. The "geocentric" model with planet Earth at the centre remained a standard until the heliocentric view of the world, with the sun in the centre, was eventually confirmed by Copernicus [20] in 1514 AD, by Galilei[21] in 1590 AD and then finally accepted as the correct order. The "heliocentric view of the world" only refers to the solar system where the planet Earth is included.

2.2.5. Up to the beginning of the 21st century, man has not yet been able to establish what light really is. To understand light was completely impossible up until the pre-scientific era of the philosophers. But even after that, it remained impossible, except for Pythagoras, whose theory was both logically and mathematically correct. It was, however, considered too improbable to be accepted completely, but it did lay the foundation for future light theory development. As Pythagoras had argued, people managed to determine that there was something peculiar about the concept of light, especially when it came to its reflection on water. About 200 years later, Aristotle explains the principle of light in a seemingly intelligent, but still incorrect, fashion. Aristotle's explanation has in fact been sacrosanct since then and is still seen as valid to this date but can now be proven invalid. Einstein also tried to understand the concept of light. His particle theory of light (1915 AD), however, was based on the misconception that light was light particles, as such things seemed to jump out of a piece of metal when heated, but which was in fact short EM waves being emitted from a piece of metal at high temperatures.

2.2.6. Light does not exist as real light. Light is in fact only an experience of light inside the eyes and brain of a living creature when EM waves from strongly heated atoms reach an eye. The rays of light always travel straight out in all directions from a light source and are reflected via matter, water and atoms of air. One reason for concluding that "rays of light" cannot be real light, is that the eyes and brain of different species experience different frequency ranges as light. In much the same way, various species' sense of hearing can pick up on different frequency bands to perceive sound. EM waves of light are not real light – and similarly, sound is not actually real sound, but silent EM energy waves that spread through the air from any matter subjected to physical damage, which is then perceived as sound when these waves reach the auditory organs of living creatures. EM energy waves, such as those that cause light and sound, are able to "go around corners" on planet Earth since the air on Earth contains reflective atoms. Images from the human visit to the moon, where there is no atmosphere with atoms that reflect and spread EM waves of light in all directions and around corners, show a sharp line between illuminated and shaded matter.

2.2.7. The explanation of light also helps to explain many other scientific mysteries. Man's awe and reverence for light have not wavered since the beginning. Light is the key of life.For a long time the sun was seen as a god because its light and energy are the actual prerequisites for life and vegetation on planet Earth. Nowadays, the human population knows that the sunis only one of several billion stars in space, or approximately 300,000,000,000,000,000,000,000 to be slightly more precise.Many resemble the sun but can be far smaller or more than athousand times bigger. At night, a few thousand stars can be spotted from planet Earth. Man has learned to know planet Earth and, through increased knowledge of physics, also now knows more about the conditions for various forms of life. Thishas created an understanding of the fundamental role of atoms,not only as nature´s building blocks, but also as a completely logical cause-and-effect system for everything, regulated by thecontinuous energy exchange of atoms that takes place through its EM waves. The atoms´ EM waves are also the communication principle for the senses of all living creatures, such as sight/light, hearing, taste, smell, touch and also communication betweenall biological forms of life and for all forms of interactionin nature. The energy from the atoms of the sun, which is transmitted through space in the form of EM waves, is not onlythe prerequisite for life and vegetation on planet Earth, butalso gives a clue to mankind for understanding nature and life's common, fundamental languages; the EM waves of atoms. All communication between living creatures, as well as with and between all biological forms of life, takes place through atomic energy exchange via EM waves. This includes the sound of the human voice, which is created by vocal track vibration and its emitted atomic EM waves, as well as EM waves emitted by strongly heated atoms in the form of light from the sun and all man-made communication technology.

2.2.8. Bioluminescence is light emitted from certain living organisms. This function of certain species – particularly fish and other marine animals, but also some animals on land led mankind to the conclusion very early on that the human eye also has some kind of illuminating function in combination with sight. In the past man used fireflies as a form of light as the body of these insects produced their own light. This type of light is caused by bacteria living inside the light organs of living creatures, particularly inside insects and other small animals. Light is then created by a biological organism producing a reaction, which in turn leads to energy being emitted   and that happens to have the same frequency as light. Many deep-sea fish species have, for example, a light organ inside their eyes allowing them to locate and attract prey. That human eyes would have some kind of light function was therefore seen as natural, a hypothesis advocated by many philosophers.

2.3. Pythagoras grasped the principle of light

2.3.1. The very first light theory based on logic.
Historic light theories indicate that most philosophers would name Pythagoras, who lived during the 6th century BCE, as the father of the first logical light theory. It was also this theory, which today is misleadingly called "emission theory", that became the most widely accepted up until the 1500s AD. It was not a real "emission theory" as such, since this would mean that the human eye would be emitting light. It was, in fact, saying that light must be assumed to be created as an experience or perception inside the eye and the brain of the viewer and would therefore not exist as light in the air. This theory is also consistent with what today can be considered true, as explained based on an understanding of EM energy waves emitted by hot atoms. Subsequent

generations did not, however, acknowledge Pythagoras' light theory in the sense that light is created as an experience in the eyes and brain, but was rather modified into a combined "intromission" and "emission" theory, i.e. light comes into the eye as well as being emitted from the eye. Eliminating the sunlight altogether was probably seen as unfeasible as it would eradicate the very core of Pythagoras' theory.

Statue of Pythagoras on the island ofSamos, Greece

2.3.2. Pythagoras interpreted the reflection of light on water as evidence that the impression of light must be created by the eyes and brain. "The glitter path/patch" is an ancient and challenging natural phenomenon, which can be seen as the most important clue to the explanation of light. The reflection of light on water is perceived by each individual observerat sunrise and sunset and will appear as a path of light heading straight towards each of them. If several people gather on a beach, stand close and look at the setting sun, everyone will see the same narrow, brightly shining path on the water heading straight towards the group. If these people spread out on the beach instead, with for example 50 metres between them, each individual will experience a personal glitter path heading straight towards only oneself; i.e., one individual path for each onlooker. It will not be possible to see anyone else's glitter path, just one's own. It therefore seems as though each person's eyes create a private light path heading towards the sun. The theory of a person's own "eye light", or of the sunlight being reflected by the eyes on the water, however, comes apartas the incoming waves of water would then be illuminated by the person's "eye light", but this does not happen. So there is not evena theoretical option of a personally perceived light reflection on the water being created by the eyes of an onlooker standing in the water or on the beach. The "glitter path" of the rising or setting sun forming on the surface water can therefore neither come from the sun nor from the eye of the onlooker. According to Pythagoras, the only possible explanation was that light is not real, but is created as an experience or perception inside a human brain.

2.3.3. In generations to come, a majority of philosophers backed Pythagoras' light theory. But as the theory was not clearly documented or described and/or was modified at a laterstage to make it less difficult to explain, less difficult to understand and less improbable, it was transformed into a combination of external light, for instance from the sun and internal light being emitted from the eye. That the eye does not only receive, but also emits or reflects light, was a theory that most philosophers advocateduntil about 1500 AD. Such a compromise, however, abandoned the core of Pythagoras' observations and conclusions – i.e., that light does not actually exist as such, but is created as a perception inside the eye and brain of living creatures, hereinafter referred to as "induction theory".

2.4. Pythagoras' light theory vs. Aristotle's

2.4.1. From the 4th century BCE, Pythagoras' "induction theory" had to face competition from Aristotle's "intromission theory". Aristotle's theory would be perceived as more logical, as it corresponds to the way man experiences natural light, i.e. that light comes into the eye from a light-source and only instraight lines. If this were the case, it might also explain Pythagoras´evening- or morning light-reflex on water, coming in low from the horizon in straight lines and being visible by each individual vieweron the water as heading straight to the eyes. As Pythagoras´ theory seemed extremely unlikely, with an impression of light created by theliving viewer´s eyes and brain, it was logical that Aristotle´s theory would win. Pythagoras' theory, that light is an experience inside theeyes and brain of a human, today appears illogical and most people have never heard of it. It has therefore been used to illustrate how primitive the earlier philosophers must have been. On the other hand, "bioluminescence", i.e. the fact that many living species do have eyesthat glow in the dark, could still be used as a strong argument for an"emission light theory".

2.4.2. Aristotle explained light in a simpler way. He assumed that light and the world worked much as it seemed to man. Aristotlemeant that rays of light are real beams emitted from the sun, which run in straight lines, fanning out in all directions, and that the light of each ray is only visible from the front, i.e. in a straight line. This was a seemingly ingenious idea, because it seemed to not only explain light reflecting on water, but also why the light heading from the sun at night and onto the moon is not visible laterally from Earth. This would mean that Pythagoras' obscure and unlikely theory was no longer needed. It also led to Aristotle's seemingly more natural theory of light becoming the official scientific theory, which is still valid to this day.

2.4.3. Aristotle´s light theory is today proven incorrect and Pythagoras´ light theory correct. It has actually been proven that the kind of light rays Aristotle imagined, with light only in a straightforward direction, do not exist in reality, for the reason that light is today understood as actually not being real light. Light is, instead, always EM waves perceived as light by the eyes and brain of a living creature. Aristotle's theory is therefore invalid, but can still be admired for creativity. His hypothesis of real light, however,remains the scientific understanding and view of the world to this day. Pythagoras' light theory, later substantially modified, was considered by most of the early philosophers to be the valid theory, a view that many prominent opticians shared for several centuries. It was, however, not easy for people of that era to logically understanda light principle and world view as diverse and complex as the one prompted by Pythagoras' theory. The problem has therefore been that as long as it was impossible to understand Pythagoras' light theory, or indeed organise it into a logical world view, it has not been meaningful, or even possible, to question or discuss it. That Pythagoras would be proven right and that light is indeed created as an experience inside the eye and brain, was extremely unlikely as long as Pythagoras' insights could not be supported with more convincing evidence.

2.4.4. Even though Aristotle was incorrect, his theory is stillthe scientifically accepted explanation of light. It must be regarded as strange that the issue about the nature of light has still, after almost 2,500 years and with modern scientific resources, not been solved and the Aristotelian theory has not been unravelled. This may be due to the fact that Aristotle's natural and seemingly logical light theory, dating back to the beginning of religious movement around the year 1000 AD, quite simply turned into the onlytheory allowed. Pythagoras' theory had probably been prevented byreligious groups in order to maintain a traditional and, as far as man is concerned, natural view of the world. Science has, therefore, had only the Aristotelian theorem to turn to as an apparently logical explanation to light and when state religions were established with their tailored history description and teachings, no change to the world view was permitted.

2.5. From the idea of real light to EM waves perceived as light

2.5.1. This question was originally raised by Pythagoras, i.e. whether light is real light as coming from the sun, orwhether light is not a substance but merely perceived as light, caused by an unknown principle according to Pythagoras' theory. Or if reallight comes into the eyes from the sun or other light sources, as per Aristotle's and today's theory. The issue of light has partly been mixed up with the visual function and whether the eyes are just passive recipients of visual information, or if they have an active function with outgoing rays to retrieve information from the viewed object. It was soon established, however, that an active outward visual principle could not be possible, given the enormous distances in space which the outward rays from the eyes would have to be transported in order to reach the stars. This distance would then have to be covered each time you open your eyes towards the night sky. Instead, it was assumed that the light would either continuously come to the eyes as real light, which became Aristotle's theory or, as per Pythagoras' theory based on the morning or evening sun's reflection on water, that light has no substance but is experienced as light inside the human eye and brain.

2.5.2. Light does not exist - light is, in fact, EM energy waves perceived as light in the eyes of a living creature. What Pythagoras understood as light being created inside the eye can today be confirmed as perfectly correct. At the time, though, he could understand neither how nor why this occurred, as EM waves were not understood. If EM waves come directly from a light source, they are perceived by the eye simply as light and if they are reflected via matter, they are perceived as an illuminated image of thismatter. Aristotle meant that the rays of light are only radiating in a straightforward direction. This is partly correct, as an observer on the beach will always experience the low morning or evening sunlight coming along a narrow path on the water, privately to each observer. Man of today can, however, conclude without any doubt that light does not exist as real light, but is always EM waves emitted from heated atoms of a light-source, such as the atoms of the sun or a light bulb. When these EM waves reach the eye and brain of a living creature, they are perceived as light.

2.5.3. Life invented light. An updated understanding of the light does not need to clash with the traditional understanding of light as "the beginning of everything". On the contrary, the creation of life with eyes and brain tailored to experience the EM rays of the sunas light, was in fact the very beginning of light. It is actually life, equipped with eyes and brain, which individually creates its impression of light by transforming the EM waves of the sun, the stars and the fire into an image experience of light. Before life with eyes and brain existed, there was no light in the universe. Light is the invention of life.

2.5.4. The real nature of light has still, at the beginning of the 21st century, not been scientifically clarified. Light has been explained by man through two different physical theories; the particle theory and the EM-wave theory. Einstein's particle theory, developed in the beginning of the 20th century, has the advantage of being mathematically more understandable and manageable, while light in the form of EM waves is the correct physical explanation. This has been convincingly proven for more than 300 years, first by Huygens [22] and then confirmed by Maxwell[23]. However, compared to EM waves, the particle theory made it much easier, or actually just possible, to describe the atoms´ energy process mathematically. The atomic particle theory was therefore necessary for practical reasons, even if the EM wave theory of light was and is the scientifically valid and accepted theory. Both theoriesare consequently scientifically accepted at the beginning of the21st century, which is troublesome, as the EM wave principle is the true communication system of atoms. The particle theory of atoms, whose most important and early supporter was Newton (1642–1727 AD), was thereafter also supported by Einstein at the beginning of the 1900s, until EM waves in the 1920s were finally proven to be the energy language of atoms. Despite this fact, the particle theory is still necessary as a "mathematical tool", which is a function EM waves do not provide. These two complementary but different functions have confusingly been called "wave-particle duality", which is not correct, as it proposes that waves and particles are equal or perform the same function. That is not the case – but they are complementary in the way that EM waves explain/represent the type of energy communicated, but not the amount; and particles explain/represent the amount of energy, but not the type.

2.5.5. EM energy waves cannot be mathematically quantifiedin the same practical way as particles. This is why the particle principle must continue to be used as the quantitative mathematical language of the atom. Unfortunately this has created scientific inconsistency and confusion in the matter, as it gives the general impression that particles, due to mathematical quantification, are regarded as the language of atoms despite the fact that this quite evidently is EM waves, which cannot be as easily quantified. It has, therefore, been chosen to solve this problem by accepting a "wave-particle duality", which would mean that EM waves andparticles are principally the same thing. This is where the final challenge starts. A wave is a wave and a particle is a particle. It is notpossible to solve the problem by proposing that they are the same. What was not understood at the time was that EM waves and particlesrepresent two completely different physical ideas of how atoms communicate and it would become apparent in the beginning of the 21st century that communication can only be achieved via EM waves.There is actually no theory that can explain how this could possibly happen with particles. Einstein's particle theory of atoms, inherited from Newton, was however the only theory to describe and quantify the atomic energy system, which was necessary for mathematical reasons. Even in the 21st century, the particle theory is still necessary, because no corresponding mathematical principle to quantify EM waves exists as of yet. The "wave-particle duality" is a creative solution in order to move forward in the process of understanding the "language of the atom", but it has to be understood that duality does not mean that wave and particle are the same and can do the same job. Duality means that "waves" are the actual language of atoms, while "particles" are their mathematical language.

2.6. The most prominent philosophers of light, with theories - in chronological order

As a background and to support the explanation of light, it is important and interesting to understand how the light problem has been discussed historically, as well as how it has been considered by eminent philosophers. This compilation represents a selection of the most famous philosophers and their light theories.

Historic theories of light:

Intromission:
Real light comes from an external light source – 16 philosophers

Induction:
The experience of light is created in the brain – 6 philosophers

Emission:
Light is emitted from the eyes in combination with external light
– 5 philosophers

2.6.1.	Pythagoras	(570–495 BCE)	Induction
2.6.2.	Empedocles	(490–430 BCE)	Induction/ Emission
2.6.3.	Leucippus	(480– ? BCE)	Intromission
	& Democritus	(460–370 BCE)	Intromission
2.6.4.	Socrates	(470–399 BCE)	Induction
2.6.5.	Plato	(428–348 BCE)	Induction/ Emission
2.6.6.	Aristotle	(384–322 BCE)	Intromission
2.6.7.	Theophrastus	(371–287 BCE)	Induction
2.6.8.	Epicurus	(341–270 BCE)	Intromission
2.6.9.	Euclid of Alexandria	(325–270 BCE)	Intromission
2.6.10.	Archimedes	(287–211 BCE)	--
2.6.11.	Ptolemy	(100–170 AD)	Intromission/ Emission
2.6.12.	Galen	(130–200 AD)	Intromission

2.6.13. Conclusion of the issue of light around 500 AD.

2.6.14. Al-Kindi	(801–873 AD)	Induction/ Emission
2.6.15. Alhazen	(965–1040 AD)	Intromission
2.6.16. Grosseteste	(1175–1253 AD)	Emission
2.6.17. Galilei	(1564–1642 AD)	--
2.6.18. Kepler	(1571–1630 AD)	Intromission
2.6.19. Gassendi	(1592–1655 AD)	Intromission/ Particles
2.6.20. Huygens	(1629–1695 AD)	Intromission/ EM waves
2.6.21. Newton	(1642–1727 AD)	Intromission/ Particles
2.6.22. Maxwell	(1831–1879 AD)	Intromission/ EM waves
2.6.23. Russell	(1872–1970 AD)	Intromission/ EM waves
2.6.24. Einstein	(1879–1955 AD)	Intromission/ Particles
2.6.25. de Broglie	(1892–1987 AD)	
& Schrödinger	(1887–1961 AD)	Intromission/ EM waves

2.6.1. Pythagoras (570–495 BCE) (Light theory: Induction)

1. Pythagoras was the first to call himself a "philosopher". Philosopher means "the one who tries to understand". This was less pretentious than the previous word fora wise man, a "sage", suggesting that healready knew everything. Pythagoras' teaching was more a way of living, with a strong social and religious message, than an attempt to explain how the world works. He was also a most prominent mathematician. It is clear that his knowledge in mathematics and science was based on an unusual intelligence. Early in life, Pythagoras had been allowed to accompany his father on long journeys and was both sophisticated and educated, not only in terms of plain subject knowledge, but also when it came to music and poetry. He was also exclusively introduced into philosophy as a private disciple of Thales and subsequently spent a long period of his life, probably between the ages of 22 and 50, in prosperous cultures like Egypt and Babylon. His time in these cultures was apparently mainly used for studying ethics and religious teachings, but also mathematics and cosmology. Pythagoras was a philosopher who advocated a way of life in combination with knowledge of the world.Pythagoras can also, during the latter part of his life, be considered the father of modern religion in the western world, combined with awide range of educational activities as part of a permanent academy.

2. Pythagoras meant that the experience of light occurs inside the eyes and brain ("Induction"). This has sometimes been interpreted as light being emitted from the eyes, but which has been mixed up with the fact that the eyes of some animals can glow in the dark. This phenomenon was also historically known because insects, fish and most animals emit light from their eyes during the night. Humans have for example used cages of fireflies as a kind of living light source. That light might also be coming from human eyes was therefore not considered an unreasonable idea and was, for a long time, accepted by most philosophers. Pythagoras evidently did not mean that eyes emit light as such, but that the experience oflight is created inside the eye and brain and that light does not exist externally in the world around us.

3. The sun's early morning and evening reflection on water – "the glitter path" – will always head directly towards the eyes of each individual onlooker.[24] People's spontaneous general thought is that the light reflection on water is created by the rising or setting sun upon the horizon and that this should naturally illuminate the whole sea. But, as a matter of fact, as the narrow glitter path of the morning and evening sun is always experienced as heading straight towards each individual onlooker, the conclusion must be that the path of light is not created by the sun itself. The only explanation can be that EM waves from a light source are not real light, but merely perceived as light by the eyes and brain of a living creature. There is reason to believe that Pythagoras had observed how the sun's narrow morning and evening reflection path on water always leads individually to each spectator, as observed by each spectator. He had probably concluded that since the reflection obviously cannot be directly caused by the sun, the experience or perception of light must instead be created inside the human eye and brain.

4. Pythagoras probably created his light theory in Greece, Egypt or Babylon. A correct understanding of light would, in fact, match Pythagoras' theory when it comes to inward EM energy waves, which are not real light, but perceived as light by the eyes and the brain. Perhaps Pythagoras had personally observed, considered and solved the matter of light reflecting on water as a youngster on the island of Samos, where he was born. Or perhaps he learnt this while he lived in Egypt or Babylon, cultures that were considerably more advanced in the scientific sense than Greece. It is likely that he had spotted the glare on the surface of the water and contemplated the logic behind this phenomenon. The same applies to mathematics, Pythagoras' main area of expertise, which he had initially learnt from his mentor, Thales, but also during his time in Egypt and Babylon, as maths was the special competence within these countries. There is no doubt that Pythagoras was an unusual mathematical genius.

5. "Pythagoras should be considered to be the most influential of all Western philosophers ever." (*A history of Western*

Philosophy – Bertrand Russell).

2.6.1. Empedocles [25] (490–430 BCE) (Light theory: Emission)

1. The first person to describe light in writing and develop a theory of vision. Born into a famous family in Sicily, blessed with a strong personality and a solid education and influenced by Pythagoras, Empedocles became a prominent philosopher and convincing speaker with great knowledge about both nature and medicine. Aristotle called him "the father of rhetoric". By mixing part philosopher, part prophet and part scientist together, as well as a strong religious faith, Empedocles resembled his role model Pythagoras to a certain extent. He probably knew of Pythagoras' light theory, as he was born five years after Pythagoras´ death. Empedocles also described Pythagoras in praising words: "amongst them was a man of incredible wisdom, an exceptional master in all areas of knowledge." Empedocles was not only interested in history and nature, but also harboured scientific theories of causality, perception and thinking skills, as well as explanations of celestial phenomena and biological processes. He knew, for instance, that the moon shines at night by reflecting light rays from the sun, which at that point is located on the other side of Earth. Apart from having been an important historical person scientifically, Empedocles is also known for his spectacular death by throwing himself into the volcano Etna.

2. Empedocles' "visual rays", later called "emission theory", claimed that the visual experience is created by the eye reflecting rays of light. This became the dominant light and visual theory for about 600 years, from ca 450 BCE to 150 AD. Empedocles' light and visual theory was a modification of Pythagoras' theory, which was a "lightless light" that humans perceive as light, similar to today´s EM waves. Empedocles meant that the external "light" is real light that comes from the sun or another light source. He also claimed that the eye reflects the light, which explains why the light reflecting on the surface of the water is a personal experience. At the time this conclusion was a perfectly logical explanation. Empedocles believed eyesight exists as a combined effect of light coming into the eyes and visual rays being emitted from the eye, therefore called the "emission theory". This could perhaps be a logical principle, if light really is light. However, Pythagoras had realised that light is not something that exists as a separate phenomenon, but is the human brain interpreting something that creates an experience of light inside the eye and the brain.

3. Empedocles' theory of real light coming into the eyes seemed natural to man. The difference between Pythagoras' theory and Empedocles' was that Pythagoras argued that light does not exist as real light, but is created as an experience inside the brain; while Empedocles explained the visual experience with a combination of the "intromission theory" and the "emission theory" and that eyesight is created by external light, predominantly the sun, reaching the eye and then reflecting and being emitted as visual rays.

2.6.1. Leucippus [26] (480 BCE–death unknown) & Democritus (460–370 BCE) (Light theory: Intromission)

1. Leucippus was the founder of a philosophical school and discovered the theoretical existence of atoms. Out of the most important philosophers in history, Leucippus is perhapsthe least known, because there are no written documents left thatcan be attributed to him. He is, however, mentioned in old referencesources as the first person to develop an atomic theory for everythingthat exists and was the founder of a philosophical school called the "atomists". The main evidence for Leucippus' existence and significance may instead be attributed to his pupil Democritus, who came to develop and promote the atomic theory and to Aristotle, who mentions Leucippus several times in his writings. It is stated that Leucippus established a school in Abdera somewhere around 440 BCE, which Democritus joined. A different city, Metapontum, was founded by Leucippus and the city later paid tribute to him by putting his image on a coin. Leucippus is often referred to as Democritus' teacher and both of them are considered to be the founders of the atomic theory, even if Leucippus was the original "atomist".

2. According to Leucippus, the world is nothing but atoms and void. Everything visible consists of atoms, which are invisible to man as solitary units. Both Aristotle and his pupil and successor Theophrastus, have called Leucippus the inventorof the atomic theory. Therefore, there is no doubt of his significance,even if Democritus is mentioned the most. As the theoretical explorer of the atom, Leucippus must in fact be considered one of the most important persons in the history of science. As far as Leucippus was concerned, the world is nothing but atoms and void. Anything and everything that exists consists of atoms.

3. Democritus meant that matter contains tiny images that are constantly being emitted and that, together with light, can be distinguished as they reach the eye. Democritus has deservedly been honoured for his "atomic doctrine", even if Leucippus was the father of the theory. Democritus lists many Greek philosophers in his publications, but stresses that Leucippus, the creator of the atomic doctrine, influenced him the most. Democritus argued that matter contains tiny images that are constantly being emitted and that, together with light, are perceived or distinguished when they reach the eye. This theory is very similar to the modern understanding of vision/sight, which means that light rays hit the surface atoms of the matter and are then reflected as images to the eye. Democritus seems to have taken over and systemised Leucippus' theories, even if some parts are considered to be written by Leucippus. The majority of text alternates between refer either to both, or only to Democritus. These two philosophers meant that everything in the universe is either empty voids or small particles, which Leucippus called atoms; a word that means indivisible. The atoms collide from time to time and form clusters. Everything can be explained by atoms and the atomic movement, including the living body and the human spirit. Democritus followed Pythagoras' example by spending several years in Babylon and Egypt, studying their mathematics and science.

2.6.4. Socrates[27] (470–399 BCE) (Light theory: Induction)

Like Pythagoras, Socrates was more interested in ethical philosophy than natural philosophy. Socrates is considered one of the founders of the ethical philosophy. He was enigmatic as a person and known only through his pupils. In the dialogues of his disciple Plato, Socrates is the main character when it comes to knowledge, theory and logic and is the one who conveys Plato's thoughts. Socrates did not personally leave any writings for posterity, but inspired many disciples, including Plato.

1. Socrates meant that eyesight occurs through cooperation between the eyes and the brain. This was a reasonable insight since it is the brain that converts the EM energy waves/rays of light to a perception of light and image when the reflected rays of light from the illuminated matter of the surroundings reach the eyes. Socrates also argued that even if the light of vision is similar to the light of the sun, they cannot be compared as sunlight is far superior. At the age of 71 Socrates was sentenced to death by poison because his philosophy was considered harmful to young people. He would have been freed if he had just taken back everything he had said he stood for, but decided to stick to his belief.

2.6.5.Plato (428–348 BCE) (Light theory: Intromission/ Induction/ Emission)

1. Plato was influenced by both Pythagoras and Socrates. Pythagoras' followers, the "Pythagoreans", are considered to have had a major impact on Plato's teachings. Plato was the founder of the Academy in Athens, the first institution of higher education in the western world. Aristotle attended as a pupil and later became Plato's successor. Plato meant that the visual experience does not represent what the world looks like, but what it may look like. This is logical because each living species experience the world differently, depending on the design of the eye and brain of that particular race or breed. Plato proposed a theory of light, which contained three different rays of light: 1. from the light source to the eye, 2. from the eye to the observed item, 3. from the observed item to the eye. Today we know that visual experience is created through EM energy waves from a light source being reflected into the eys by the surface atoms of the matter. This principle of light is consistent with the light theory of Empedocles (490–430 BCE), influenced by "Pythagoreans", with the addition that the rays of light are refracted from the observed item to the eye.

2. Plato considered light to represent the good. Plato used the light of the sun as a metaphor for enlightenment and to describe the full sense of the good light that offers its light, so that man shall be able to see and learn to understand the real world. Plato was also famous for his allegories, i.e. symbolic depictions, which for more than 2,000 years became a popular way to express oneself that lasted until the 18th century. His praise of the sun and the light is representative of Plato's imagery, e.g. "The good light is the basis of all truth"; "Without the good light, we would only be able to see with our physical eyes and not with the eyes of our soul"; "The sun leaves light as its legacy"; and "Good light sheds its light so that man may experience the true reality". Plato's philosophical influence on the medieval way of thinking and expressing oneself was extremely significant.

2.6.6. Aristotle (384–322 BCE) (Light theory: Intromission)

1. At the age of 18, Aristotle started teaching at Plato's Academy in Athens. He was there for 19 years, until Plato's death (347 BCE). After Plato's death, Aristotle changed his philosophical beliefs from Platonism to empiricism, which meant a shift from belief based on theory, to belief based on logic and evidence. Aristotle was a systematician. At 41 years of age (343 BCE), he was the private teacher of Alexander the Great[28], who was 13 years old at the time and this arrangement lasted for about three years. As King of Macedonia, Alexander the Great came to conquer large parts of the "Eastern" world; from what is today Greece/ Macedonia, all the way to the Asia Pacific region. With donations from Alexander the Great, Aristotle created a library and academy in Athens, called the Lyceum, with hundreds of books on natural history, as well as created the first-ever zoo and botanical garden. The Lyceum focused more on empirical natural science, unlike the Plato academy with its idealism, mathematics and speculation. Aristotle's scholars were given the responsibility of setting up and running the library. His close relationship with

Alexander the Great meant that zoologists, botanists and others connected to the academy could join royal expeditions. During the years of 335 to 323 BCE, Aristotle and his students made important scientific discoveries in a wide range of subjects such as anatomy, astronomy, embryology, geography, geology, meteorology, physics, zoology, philosophy, logic, aesthetics, ethics, political science, metaphysics, politics, economy, psychology, rhetoric, theology, education, literature and poetry – knowledge that laid the foundation for western philosophy and science.

2. The strong connection to Alexander the Great ensured Aristotle had significant funds for scientific studies. Aristotle's extensive work with systematically categorising the world of plants was indeed an epic and important project, but did not contain any ingenious component, as such. The inability to grasp Pythagoras' interpretation of light, i.e. that it is created inside the human brain, could be seen as one example. He was, however, extremely knowledgeable, a systematician and a pioneer of organised research at Lyceum, his education and research centre. One can assume that the large variety of books that have been attributed to Aristotle and which deal with so many different topics, are not necessarilyall written by Aristotle himself, but that he probably received help from his pupils. No individual could be expected to create such a vast quantity of scientific work in so many different areas. Alexanderthe Great died in 323 BCE, wherefore Aristotle was forced to flee from Athens and died the year after, in 322 BCE, at the age of 62.

3. Aristotle's "intromission" theory created a false concept that became the contemporary understanding of light. Aristotle's "intromission theory", that light comes as light through the air and into the eyes, as well as a seemingly logical explanation of light and vision, became the contemporary principle of light. Thiswas based on rays of light only being light at length ends, which was Empedocles' theory. This would explain why the light fromthe sun, that makes the moon shine at night, is not visible from the side when it passes Earth. Today it is obvious that light rays are not actually light at all, not even longitudinally, but that the eye and brain perceive them as light because in a straight-forward direction, they convey a particular frequency of energy in the form of EM waves from highly heated atoms, e.g. from the sun.

Aristotle's light theory, which originated in attempts to rebut Pythagoras' light theory, is therefore incorrect and invalid as disproof.

4. Aristotle is said to be the founder of logic, but Pythagoras was right about light. Aristotle's logic was based on the principle that reality is exactly how man perceives it. His main argument was that light cannot be seen to radiate from the light like a "visual ray",as per Empedocles (page 37), but only from a light source. Nor did Pythagoras mean that light originates in the eye, but that it is createdas an inner experience inside the eye and the brain. As Pythagoras himself did not leave any written documentation, the scientific theory of light started with Aristotle's seemingly logical interpretation, i.e. rays in the air, which is only light in the direction of the ray. In reality, this phenomenon is not rays of light being emitted from highly heated atoms, but EM energy waves that are perceived as light when they reach the eyes of a living creature. It is by the conveyed continuous energy in these EM waves of a light source, e.g. the sun, light bulbs or fire, that the eye and brain of the living creature create the perception of light. Pythagoras was therefore correct in his conclusion that light is created inside the eye and brain, i.e. when the energy waves from alight source hit the eye.

2.6.7. Theophrastus[29] (371–287 BCE) (Light theory: Induction)

1. Theophrastus was a disciple of both Plato and Aristotle. He became the successor to Aristotle in his creation Lyceum and continued the tradition of exploring philosophical and scientific theories. There is information attributed to Theophrastus, also called"the father of botanics", which states that "the eye has built-in light".He obviously did not accept, at least not fully, Aristotle's theoryof light coming from the air and into the eye. At the same time, it indicates insight into the correct understanding of Pythagoras, that light does not exist as a real substance in the air, but is created as an experience, a perception, in the eye and the brain, which can be interpreted and described as light being emitted from the eye.

2. The first person to try to analyse and understand light and colours. When it came to systemising the understanding of na- ture, in the spirit of Aristotle, Theophrastus became the first person to debate the issue of light and colours in writing. He left approximately 200 literary works with philosophical, ethical and scientific content and is together with Aristotle classed as the pioneer or founder of systematic botany. He described light and colour as follows:

"We never see a color in absolute purity, it is always blended, if not with another color, then with rays of light or with shadows and so it assumes a new tint. Air and water when pure are by nature white and fire and the sun yellow. Earth is naturally white. Darkness is due to privation of light. Light is clearly the color of fire; for it is never found with any other hue than this and it alone is visible in its own right, whilst all other things are rendered visible by it. But there is this point to be considered, that some things, though they are not in their nature fire nor any species of fire, yet seem to produce light. So we cannot say in the same breath that 'The color of fire is identical with light and yet light is the color of other things besides fire,' but we can say, 'This hue is to be found in other things besides fire and yet light is the color of fire'. Anyhow, it is only by aid of light that fire is rendered visible, just as all other objects are made visible by the appearance of their color".

2.6.8. Epicurus[30] (341–270 BCE) (Light theory: Intromission)

1. Epicurus, like Aristotle, advocated a scientific method. This method is based on the idea that nothing is credible if it cannotbe proved by means of observation and logic. Just as the generation before, Epicurus was in favour of the seemingly logical "intromission theory", which argues that light exists and comes externally from the air and into the eye. He was also a pioneer for the scientific method, where nothing would be accepted if it is not both observable and logical to understand. Hardly any of Epicurus' 300 works have been preserved. Epicurus, therefore, cannot be proven to have promoted the "intromission theory", apart from what has been mentioned in historical documents. His philosophy was based on natural functions being logical and true in the way that human senses perceive them, thus matching Aristotle's claims. He also stressed exemplary friendliness as an important component andfounded a school of science based on this principle.

2. Epicurus was an atomist, same as Leucippus and Democritus and believed that everything consists of small pieces of matter. Epicurus too explained that everything in nature interacts with the sense organs, which is a historically important observation and theory, but does not seem to have gained scientific acceptance. Vision and the visible were also explained as synergy between atoms and our eyesight. He stated that all matter constantly sends out "eidola" through the air and when they reach the eye, they describe the objects that they come from. This principle is, in fact, very similar to the current understanding of EM waves of light, even if it does not use the same terms as we do now. Epicurus also explains the other senses in a similar way, such as the idea that flavours and taste are created by special atoms making contact with the tongue. This explanation, offered more than 2,300 years ago, resembles our modern understanding. What Epicurus called "eidola" is today known as electromagnetic waves (EM waves).

2.6.9. Euclid[31] (325–270 BCE) (Light theory: Intromission)

1. "The father of geometry and optics". From 300 BCE onward, optics was developed as the science of light and vision. Euclid of Alexandria was the first to base the study of vision on mathematical principles and taught students like Archimedes, who was both a technical and a mathematical genius. Around the year 300 BCE, Euclid wrote the first of many works, titled "Optica", which contains a summary of the characteristics of light, but mainly deals with the geometry of vision. In the 16th century, his book "Elements" became the most printed book ever next tothe Bible and is considered to be the most influential textbook ever written. The content of the book seems to be mainly supported by material from Pythagoras' mathematics, which underlines Pythagoras´ historic importance as a mathematician.

2. Initially Euclid followed Plato's theory of vision, being rays emitted from the eyes. I.e., according to the "emission theory". Plato's theory was considered to be more logical than Pythagoras' original light theory, where light was said to be created as an experience inside the brain. Euclid later questioned Plato´s theory that visual rays could be emitted from the eyes, as the eyesight reaches the stars immediately when the eyes are open tothe starry sky. He meant that no rays from the eye can reach a star that quickly. Something would have to come from the star instead. What actually happens is that the EM energy waves from the atoms of a star permanently reach earth and can immediately be utilized in return by people to see the stars. This is the same way EM waves operate on earth – continuously emitted from all kind of sources andreceived by living creatures, except when sleeping.

These receivers are all equipped with five senses in order to be able to translate the information received, which is always in the form of EM waves, the language of atoms. Information is always being emitted or provided from a source, never the receiver fetching it.

3. Euclid changed his belief from "emission theory" to "intromission theory". He now meant that rays of light must come from an external light source and into the eyes, compared to his earlier theory of the rays of light being emitted from the eyes. Euclid's changed position adhered to Aristotle's natural and logical light theory, which was accepted by most people. This was a strategically sensible choice, in view of the enormous scientific success and fame Euclid later achieved with his geometry. Had he persisted and stuck to the emission theory, he would have been seen as controversial and would have faced the challenge of proving it. As the father of geometry and mathematics, it was not worth the risk for Euclid to jeopardise his authority by having to defend the "emission theory". This, especially, as Aristotle had already convinced the scientific world that the "intromission theory" of light wascorrect. As at that time the concept of EM waves was not yet known, the real explanation of light was not possible to grasp and understand.

4. Euclid of Alexandria is not to be confused with Euclid of Megara. Euclid of Megara was another prominent philosopher, who lived approximately 100 years before the more famous mathematician Euclid of Alexandria. Euclid of Megara was one of Socrates' students and the founder of the Philosophical School of Megara, where he taught logic. His school lasted for moreor less 100 years.

2.6.10. Archimedes (287–212 BCE)

1. Light as a weapon of war.[30] Archimedes is considered the most prominent combination of mathematician, physicist, engineer, astronomer and inventor of all times. His unusual demonstration as a contribution to man's understanding of lightwas based on advanced knowledge of optics, which was available atthe time through Euclid, who had been Archimedes' tutor in Alexandria and had laid the foundation of optics as a science. Since then, it has developed into an important branch of science, both for the human needs of maintaining eyesight and the ability to see small objects with the help of instruments, or for long-distanceobservations.

2. The first person who tried to measure the speed of light and who advocated the "heliocentric" view of the world. I.e., with the sun in the centre. Archimedes concluded that light moves so fast, that it is impossible to time it as such. He admitted he didnot understand what light really is, but appreciated it had developedinto an important branch of science, both for the human needs of maintaining eyesight and the ability to see small objects with the help of instruments, or long-distance observations. Archimedes' insights and important contributions to applied optics, as well as mechanics, actually comprise one of the most important chapters in the history of science. Just as light was once the beginning of all life forms with eyes, Archimedes' creative understanding of light became the start of human technological development, largely in the area of weapons. One of Archimedes' methods of war involved meeting the enemy fleet with reflected sunlight from large, shiny, parabolic mirrors of bronze or copper, whose rays were so hot they set ships and sailson fire and burnt the crew. A written document from the siege of Syracuse (214–212 BCE) describes how Archimedes' technique with concentrated sunbeams destroyed enemy ships and in other documents describes Archimedes' weapons as "burning glass". Thistechnique has been called "Archimedes' heat ray". He constructed other weaponry too, such as catapults and stone throwers.

3. "Archimedes' heat ray" was tested at a naval base in Greece in 1973. Seventy copper-coated discs, 1 x 1.5 m in size, were directed at a wooden ship approximately 50 metres away and as a result the ship caught fire within just a few seconds. This was clearly a very effective and incredibly nasty weapon. Several similar tests have been performed and all results have been similar.

2.6.11. Ptolemy[31] (100–170 AD) (Light theory: Combinationof emission / intromission)

1. Ptolemy was a Greek mathematician, music theorist, astronomer, geographer, astrologist and poet. Just like Pythagoras, he had extensive knowledge in musicology based on the relationship between music and mathematics. Ptolemy describedhow musical notations actually represent mathematical equations, which has been called "Pythagorean tuning" as it was originally discovered by Pythagoras. Ptolemy describes light from different aspects, which was fundamental to light science.

2. Suggested a combined version of the "intromission" and the "emission theory" of light. Ptolemy advocated the combined principle of light with both external light from the sun and visual rays from the eye, which gathers information about nature and distances. Like most theorists of the past, he assumed eyesight is created by something emitted from the eye and, according to Empedocles' original emission theory dating back to 450 BCE, was called "visual rays". This was based on comprehensive knowledge of the physiology of the eye and mathematical considerations. Distanceassessment seemed to be a particularly good reason for assuming the existence of a combined "intromission" and "emission" theory of light. At the time, it was not possible to understand the real principleof eyesight, i.e. incoming EM waves to the eye from a light source, directly or reflected, entering the eye and conveying eyesight and distance assessment in return.

3. Ptolemy adhered to Pythagoras' theory to the extent he felt was logically feasible. The difference was that Ptolemy, likeall other philosophers, could not believe that the energy of the sun was nothing but a perception of light created inside the brain of a human, which Pythagoras had realized issue on water. Ptolemy studied light from a range of geometric aspects. Apart from intromission of "natural" external sunlight into the eye, he suggested that combined visual rays and rays of light are emitted from the eye onto the observed object, which in return reports on distance and appearance to the eyes and brain. To a certain extent, it resembles the real principle with EM waves of light that, when reflected off matter and reaching the eyes, create a light and image perception of the observed object.

2.6.12. Galen[32] (130–200 AD) (Light theory: Intromission)

1. Galen's philosophy was in principle the same as Aristotle's and Epicurus'. Galen had extensive education upon which he could base his successful career as a doctor and philosopher andthat could be compared to the natural philosophy of Aristotle and Epicurus. During the 2nd century, Galen advocated the intromission theory, which due to his medical authority came to have a great impact on the light issue in Europe during the first millennium. This positioning contradicted Ptolemy, who believed that the visual rays are emitted from the eye and onto the observed object and then, together with rays of light, are reflected into the eye. Through his medical authority and his writings, Galen influencedthe majority of Islamic scholars to believe in the "intromission theory" , for instance Al-Kindi, whose conclusion was based on sunlightbeing harmful to the eye, to name but one. Galen based his theories onanatomical observations from the dissection of eyes. The eye was of special interest to medieval Islamic medicine and philosophy, which increased Galen's influence further and made many Islamic scholars accept the "intromission theory".

2. His acceptance of the "intromission theory" meant acceptance for the theory that seems natural to people of today, i.e. light coming into the eye. Galen's reputation asboth doctor and philosopher was legendary and was supported by approximately 500 written theses he had authored. He also travelledfar and wide to study different medical theories before he settled down in Rome, where he was integrated as a prominent citizen and later became the Roman emperor's own doctor. Galen's theories came toinfluence western medical science for over 1,300 years.

2.6.13. Summary of the light issue, approx. 500 AD

1. From around 500 AD, mankind was getting more involved in religion and social order. The debate whether light, in relation to eyesight, comes only from the air or is also reflected from the human eye in the form of visual rays, had been going on between philosophers from around the year 500 BCE. It did not cease until a society ruled by religion took care of science in present-day Europe around the year 500 AD when the medieval era, or the religious era,began. It was then decided that light came from the air, as Aristotle had argued and as people perceived to be the natural explanation. Optic science was then restricted, as light, just like the view of the world in general, should be explained according to the teachings of the church. Pythagoras' correct theory had ceased to exist. Most of what was deviant, which the early philosophers had documented, was now destroyed and knowledge academies were abolished.

2. In the year of 529 AD, Emperor Justinian closed Plato's academy in Athens, which had been founded in 387 BCE. Philosophy and natural science's debates over the major issue were then paused in Europe for a period of more than a thousand years, until Galileo Galilei, mathematician and natural philosopher (1564–1642 AD), stated that the sun was, in fact, the centre of the solar system and not planet Earth as had previously been believed. He was sentenced to life under house arrest in 1633 for this proclamation. Since then, freedom of speech and thought have progressed and developed in a positive sense, but in parts of the world it is still restricted and subject to religious traditions. In Europe, Aristotle's view of the world has mostly come to prevail, but with corrections and approval from the church. As the "father of logic", Aristotle had a simple view of science, which in the case of light meant that light in principle exists just as it is perceived by human beings.

2.6.14. Al-Kindi[33] (801–873 AD) (Light theory: Induction/Emission)

1. Al-Kindi was the first of the Islamic philosophers and like Pythagoras an advocate for the "induction theory". Al-Kindi is one of the most prominent of all scholars who have dealt with the question of light and perhaps the only one whoreally understood Pythagoras´ light theory. Al-Kindi was, just like Pythagoras, a mathematician, physicist and musician and is also known for fifteen theses on musicology with discussions about the therapeutic value of music. He was highly acclaimed and seen as the father of Islamic and Arabic philosophy by introducing and adapting Greek philosophy. Al-Kindi had no optic theories of his own, but studied most of Euclid's and Aristotle's theories; whether or not light comes into, or is emitted, from the eye. Al-Kindi's evaluation, with precise geometrical descriptions, was based mainlyon how believable the explanations of the theories were and a whole range of criteria was taken into account. He put forward the theory that everything emits rays in all directions. This is also true, as rays – or rather EM waves – from the sun are reflected by matter in all directions.

2. Al-Kindi concluded that light could only be explained by an "induction theory" i.e. that light must be an experience, a perception, inside the brain. Al-Kindi sharedthis conclusion with Pythagoras and Euclid, even though Euclid officially later changed his view to an "intromission theory", probably to adjust and facilitate his teaching career. This meant that Al-Kindi's mathematically corroborated theory was left as sole support for Pythagoras' induction theory.

3. Al-Kindi later changed his mind from the "induction theory" and instead started advocating the "intromission theory" of light. In the middle of the ninth century, the "House of Wisdom" had the largest number of books in the world. Al-Kindi's knowledge of Greek philosophy had a significant impact on his development and resulted in hundreds of theses and dissertations on various topics, in addition to light. Metaphysics, ethics, logic, psychology, medicine, mathematics, astronomy, astrology, optics and a variety of other subjects including zoology, tides, mirrors, meteorology and earthquakes, to name but afew. Al-Kindi subsequently amended his theory of light to eventually advocate the intromission theory, as Aristotle and Euclid had done before him. After Al-Kindi's death, his philosophical books sunk into oblivion and most of them disappeared altogether. His scientificlegacy was taken on by Alhazen, also called Ibn al-Haytham, approximately 200 years later, who would be the first to explain eyesight correctly.

2.6.15. Alhazen[34] (965–1040 AD) (Light theory: Intromission)

1. Alhazen, or Ibn al-Haytham, was like most other people in favour of the "intromission theory" of light. His argument was that eyesight is created by light coming into the eye and not theeye emitting light. He also gave a lot of evidence to back this up. Alhazen was a physicist, mathematician and astronomer. He is considered to be the first theoretical physicist and also wrote an influential "Book of Optics" in seven volumes. It does not only dealwith the physical theory of light, but also the anatomy of the eyeas a recipient of an endless amount of light rays that according to Aristotle are reflected from every point in the surrounding matter. Alhazen was, like Aristotle, advocating the "intromission theory" and produced a comprehensive and systematic analysis of the Greektheories available at the time. The book also lists arguments against the "emission theory", particularly against Ptolemy's theory of rays radiating from the eye and argued the theory of light being reflected in straight lines from all surrounding matter and into the eye.

2. Alhazen meant it was unfeasible to think that the eye could spread its light to the stars at exactly the samemoment the eye opened. The conclusion must therefore be that the "intromission theory" applies as the light comes from an external light source and into the eye, either directly or reflectedby the surrounding matter. His account and conclusions are today seen as the most relevant support for a general evaluation of light theories. Alhazen's theory is a combination of Euclid's and Aristotle's "intromission theories" and was carefully explained in his "Book of Optics". As long as no one had any knowledge of atoms and their EM waves, it was not possible to create a correct theory of light. Alhazen's theory was generally ignored in the Arabic world, but was translated into Latin around 1200 AD and became a standard book in optics in Europe up until the 17th century, when Newton's light theory took over.

2.6.16. Grosseteste[35] (1175–1253 AD) (Light theory: Emission)

1. Grosseteste covered a variety of scientific topics and discussed different aspects of the light issue at the beginning of the medieval university. He was an English bishop, theologian, natural philosopher and inventor of lenses and magnifying glasses during the 13th century. He promoted the light emission theory which was considered to be backed by geometry. His light cosmology is described as a "big-bang" theory in combination with a scientific view, where the starting point was the biblical concept of "let there be light". Grosseteste saw the birth of the universe as an explosion and crystallisation of matter which thenformed stars and planets.

2. Author of vast quantities of scientific literature, including "On the metaphysics of light". Grosseteste was neither a physicist nor had any science education. Instead, he was a genuine philosophical thinker and combined this with his ecclesiastical duties. He stated that mathematics was the highest of all sciences and light was the basis of all sciences. To support this he emphasised that light was the foundation of everything and as light could only be explained mathematically, mathematics had to be the ultimate science. Grosseteste was light personified and was hailed as a patron saint at his death.

2.6.17. Galilei (1564–1642 AD)

1. The first person who tried to measure the speed oflight. He also advocated the "heliocentric" view of the world and concluded that light moves so fast that it is impossible to time it.He admitted he did not understand what light really is but assumedit had something to do with atoms. This was an insightful hypothesis as light is EM waves emitted from highly heated atoms, a fact which was not yet known to man at this point in time. The "geocentric" view of the world, where everything is centred around Earth, was the chosen theory by the church and a cornerstone of their overall knowledge and power. In 1590 AD Galilei confirmed Copernicus' model of "The Universe" from 1543, which contradicted the belief of the church and stated that Earth orbitsthe sun according to the "heliocentric theory" and not the otherway around. This was a serious challenge to the teachings and authority of the church. Recognition of the new view of the world, with the sun at the centre, made Galilei into a pioneer of modern science. It was, however, not just Copernicus and Galilei who had realised the fact that Earth and the other planets revolve aroundthe sun. This discovery was originally made by the Greek philosopher Aristarchus (310–230 BCE) but has never been given much attention or recognition.

2. Galilei was convinced that the laws of nature were not as complicated as believed and that this could be proven through experiments. At the time everything in the Bible was seen as true which Galilei vehemently opposed. The church viewed this as a serious crime and in 1633 Galilei was sentenced to lifelonghouse arrest, which in the end meant nine years. Galilei remained active in his thoughts and, for instance, proposed a method of how to measure the speed of light. However, it was not possible to time anything outside a distance of 1.5 km. He concluded that the time period was shorter than what could be measured. After the deathof Galilei in 1642 the experiments continued and in 1676 AD, Danish astronomer Ole Römer measured the speed of light between the sun and Earth and ended up with 7 to 8 minutes. The official time is 8 minutes and 19 seconds, which means a light speed of approximately 200,000 km/s. This can be compared with the calculation of Huygens in the 18th century of 220,000 km/s, Fizeau in the 19th century with 315,000 km/s and Foucault in the 19th century with 298,000 km/s. The official speed of light was established in 1975 being 299,792,458 m/s, i.e. approximately 300,000 km/s. Galilei's objectives of being able to measure the speed of light were therefore achieved. He had, however, no theory about what light is.

3. Today it is understood and confirmed that the idea of "the speed of light" does not mean a continuous light propagation. The speed of light is actually only the initial propagation of EM waves from its source. Depending on the energyof the source, its EM waves will propagate a corresponding distanceand are then temporarily established as stationary EM waves until the energy of the light source is changed or eliminated.

2.6.18. *Kepler[36] (1571–1630 AD) (Light theory: Intromission)*

1. Kepler's main physical interests were colour- and light phenomena. Kepler was like most prominent philosophers and scientists a mathematical genius. During his younger years he entertained guests at his grandfather's inn by impressing them with his phenomenal head for numbers. He was influenced by Alhazen and spent a lot of time on optics, with the reflection of light against flat and curved mirrors. Kepler argued that light is emitted from its source in an indefinite number of straight rays and that colours are perceived by rays of light being reflected on different surfaces. Light in the form of straight rays was proposed approximately 2,000 years earlier by Aristotle in his intromission theory. All light theories of that time struggled with the same issue as indeed, even today, that knowledge is still lacking with regard to lightless EM waves, which the eye and brain of a living creature perceive as light.

2.6.19. Gassendi[37] (1592–1655 AD) (Light theory: Intromission/Particles)

1. Gassendi was a French philosopher, priest, astronomer, mathematician and one of the most important people in the history of science. However, this has not really come to lightin historical descriptions. Why his life's work has not been reviewedand promoted more in modern times is still very much unknown, but might be due to his official position as a priest who at the same time held a profound scientific interest, as he was the first atomic research scientist who wanted to understand the world both physically and objectively. Gassendi's main argument was that different atomic combinations create all sorts of physical, chemical and biological structures. Another of his theories was that certain atoms have more internal activity than others, which in turn means different degrees of energy. Gassendi's atomic theory led to the conviction that the movements of stars and planets are determined by their weight relations; today known as energy relations.

2. As a 20-year-old, Gassendi was teaching both theology and philosophy. He was also active in the space science area asthe leader of a group of free-thinking intellectuals. Gassendi's interest in celestial phenomena began when he discovered and experienced a colourful phenomenon in the night sky, which in 1619 Galileo had called "aurora borealis" or "northern lights". In Gassendi's younger years, he was sent to a special school for physicalstudies, where for instance he had to try to measure the speed of soundwith the help of a cannon blast, repeat some of Galileo's physical testing and study astronomy with the movement of celestial bodies. He was profoundly interested in Epicurus' logic, physics and ethics and he also studied all forms of Greek philosophy and translated Greek documents. His translation of Diogenes Laertius' book about Epicurus became the start of his authorship. Gassendi's own book about the life and deeds of Epicurus was the most time-consuming assignment of his career.

3. Gassendi's light theory was the "intromission theory". As far as he was concerned, eyesight was a function of atomiclight, or atomic images, that reach the human eye and visual sense.This is basically the same theory that Kepler, 21 years Gassendi'ssenior, put forward, with the only difference being that Kepler spokeof light as rays and Gassendi as particles. As the image and colourperception of nature later proved to be caused by light reflecting from the surface atoms of matter, the theory was principally correct.

Gassendi's assumption of light being a stream of particles was taken over by Newton around 1670 AD and competed successfully with an EM wave theory advocated by Hooke and Huygens. At the beginning of the 20th century, the atomic particle theory was accepted as the valid light theory, particularly by Einstein, before the EM wave theory takes over again in the 21st century, but still with the particle theory as a practical and necessary mathematical principle.

2.6.20. Huygens (1629–1695 AD) (Light theory: Intromission/EM waves)

1. Huygens was the first to prove that light is EM waves. Huygens' "Treatise on Light"[38], published in 1678, proved thatEM waves are the explanation of light. In contrast to Newton's contemporary theory, which argues that light is particles, Huygens suggested that light is, in fact, EM waves, an extremely advanced insight at the time, which can be considered one of the most important conclusions in the history of science. However, it did take a long time before Huygens' theory was reasonably accepted, because people felt more confident in Newton's particle theory, as he was more famous than Huygens.

2. In 1801, Thomas Young (1773–1829) convincingly confirmed Huygens' wave theory. Young proved that light can only be explained if every individual particle emitted by a light source also sends out spherical waves (EM waves). There was no reason not to trust and rely on Young's conclusions, as at the age of 13 he already spoke 13 languages and made several scientific breakthroughs during his lifetime. Young said that out of all his scientific efforts, the confirmation of Huygens' EM wave theory was by far the most important. After he passed away at a relatively young age, hisobituary and legacy became "the man who knew it all".

3. In 1817, The French Academy of Science announced a light-theory competition. Most members supported Newton's particle theory, but when Thomas Young and the famous physician Fresnel presented a detailed report of Huygens' wave theory with mathematical proof, most of them were convinced that the wave theory had to be the truth. This still did not help. Because of Newton (1642–1727) being the greater authority, the world of science chose to support the theory where light was described as particles due to thefact that the theory was more manageable and easier to understand. When the existence of atoms was proven in the beginning of the 20th century and many saw particles to be not only the language of atoms, but also the explanation of light, the scientific world decidedto ignore the convincing proof of light actually being EM waves.It now became more important to try to understand the energyprinciple of atoms and for this, a particle theory was a more descriptive, practical and faster route than trying to understand and then in some unknown way at this point, quantify EM waves.

4. From the 1930s onward, reality caught up - at least partially - with the practical and theoretical advantages of particle physics. Since then, both the atomic particle theory and the EM wave theory have been viewed as scientifically viable and applicable. In reality, the only true atom theory is the wave theory.Admittedly, at the beginning of the 20th century, Einstein's atomic particle theory was a necessary, practical principle needed in order to describe or quantify atomic energy. However, light is impossible to explain as long as one assumes that light is particles.

5. Huygens' explanation of light is the true explanation. All human experiences of light and all forms of communication are created by EM waves. The EM wave theory for light is actuallyone of the most important insights of science, because it simultaneously confirms that EM waves are the common principle for all sensory experiences of humans and animals. As Newtondid not confirm the EM wave being the explanation of light in the 18th century, Huygens' original explanation was more and more disregarded and forgotten. Huygens' EM wave theory, dated 1678, was finally confirmed by Maxwell's insight into the existence ofEM waves in 1873[39]. From the beginning of the 20th century,there was again support for Einstein's particle theory of light and after 1930, EM waves had once more reached an equal scientific status as that of the particle theory, which is in fact a completely unnecessary compromise, as particles do not even exist. Only EM waves can be the principle for all communication betweenatoms and thus between everything in the human world. The problem is that EM waves cannot be quantified. The particle principle is therefore necessary for human understanding of physics,both in a practical sense as well as mathematically, but it does not have a place in the real world.

2.6.21. Newton (1642–1727 AD)
(Light theory: Intromission / Particles)

1. Newton brought light research to a new level. He stated that white light, the sunlight, is a mixture of all colours and that light is particles. He also researched forces such as gravity and planetary movements and eventually pursued the area of space optics by constructing an advanced reflecting telescope so that he could study planetary movements.

2. The mathematical principles of natural science made Newton into one of the most important people in the history of science. Newton's light theory of 1672, which was basedon light as particles, was presented just before Huygens' EM wave theory for light (1678). Despite the fact that Huygens' wave theory was clearly correct, it could not compete with Newton's particle theory, as Newton was a greater authority within science. Newton meant that light had to consist of particles, because lightis deflected by mass according to the gravity principle. Einstein did, in fact, draw the same incorrect conclusion approximately 200 yearslater. What we know today is that both mass and EM waves are energy and that particles, therefore, do not exist, but this principle still has to be used as a base for mathematical calculations.

2.6.22. Maxwell (1831–1879 AD) **(Light theory: Intromission/EM waves)**

1. In 1864, Maxwell proved that light is

EM waves. He meant that light is a combination of electrical and magnetic fields, that travel together like an EM wave. What he did not know was why, how or where this combination occurs. Today, we know that EM waves are energy and also the language of atoms. TheEM waves emitted by atoms actually create all sensory experiences of living creatures, i.e. not only EM waves that theeyes perceive as direct light and reflected light, but also EM waves that the senses pick up as sound, taste, touch and scent. The area visible to man within the EM field is limited to the wavelength area of 400 to 700 nanometers. The most obvious example of atoms emitting energy in the form of EM waves is the energy emitted from the highly heated atoms of the sun, which the human eye and brain perceive as light upon planet Earth. The light particles that Einstein later assumed jumped out of metal when it was heated and which subsequently led to the idea of atoms containing particles, was in fact short-term EM waves, which were incorrectly interpreted as particles. The particle theory of atoms was therefore a misunderstanding.

2. Before Maxwell's EM light theory, dated 1864, light was seen as particles as per Newton's theory of 1672. As early as the beginning of the 19th century, Oerstedt had become interested in the phenomenon of electromagnetism, whereupon Faraday observed that magnetism affects rays of light and that thesetwo phenomena were thereby proven to be linked. When Faraday, back in 1821, invented an engine that converted electromagnetic energy into mechanical energy, the electromagnetic theory was practically proven. In 1864 Maxwell compiled the observations of light into mathematical equations, which came to lay the foundation for all electromagnetic theories. Maxwell's understanding that light isEM waves was not, however, scientifically accepted until 1887 whenthe existence of EM fields was confirmed by von Helmholtz and Hertz.

3. Electromagnetism, electricity and radio waves havecreated a whole new world of communication through EM waves. Maxwell's conclusion and evidence that light is EM waves was a breakthrough for the understanding of light, but didnot mean a complete explanation as they still did not know what EM waves really are and where they come from. However, the most important thing was that man now understood how EM waves could be created and used in a technical sense. In fact, this theoretical insight into EM waves and their communication principles in space was one of man's most revolutionary scientific discoveries ever. The scientific background and explanation of why light is EM waves, why EM waves exist and how they are created inside the atom, however, have not yet been divulged or explained.

4. EM waves make up a natural principle for all types of wireless communication on Earth and in the universe. EM waves also explain the principle of human voice communication and hearing reception, as well as the other sensory experiences such as sight, taste, smell and touch. Maxwell's construction of EM waves is one of the most groundbreaking technical insights in history, asit became the premise and starting point for the communication systems of the modern world. The principle of EM waves is still not fully understood, however, because it is not yet generally knownhow natural EM waves are created. Maxwell's EM waves were,in fact, the most important piece of the jigsaw puzzle in order to completely understand atomic EM waves as the communication principle of everything. Or, as Maxwell expressed it:

"We have strong reason to conclude that light

itself - including radiant heat and other radiation, if any - is an electromagnetic disturbance in the form of waves propagated through the electromagnetic field according to electromagnetic laws."

5. Prior to Maxwell's discovery in 1864, the physical worldwas said to consist of particles and after Maxwell, it was considered to be EM waves. Einstein later led everyone back tothe theory of particles, which further on after Einstein was changed to a combination of both particles and EM waves. Maxwell realised how EM waves could be technically created by man and utilised for remote communication, but he had no knowledge or theory of how they exist naturally. Even today, in the early part of the 21st century,it is still not generally known why EM waves exist naturally and howthey come about.

2.6.23. Russell[40]

(1872–1970 AD) (Light theory: Intromission/EM waves)

1. Bertrand Russell was a philosopher in the classic sense and explored human interaction with nature. In 1912, light and vision were in the world of philosophy considered "sense data" or sensory perceptions. Sense data was considered an unknown mediating principle of communication between the atoms of everything and the human brain. The subject was dealt with in a particular area, namely perception philosophy, which discussed theories that were considered to be unexplainable from a physical perspective. One of these theories, published by Bertrand Russell, suggested that EM waves transmit "sense data" from all matter and that this principle applies to all sensory experiences. Russell's conclusion about EM waves as a communication principle for sense data matches the description, which today can be seen as the explanation of light. EM waves are in fact the principle of all communication, in nature as well as between living creatures. Bertrand Russellexplained in great detail both the principle of light and the principle that can be considered applicable to the rest of the sensoryexperiences of humans. Since Bertrand Russell's explanation of the principle of the senses, dated 1912, was based on EM waves and submitted a few years after Einstein's particle theory of both light and atoms – which had already been accepted by the scientific world - Russell's explanation of light as EM waves was no longer relevant,even though it was correct:

"Now this something, which all of us who are not blind know,is not really to be found in the outer world; it is something caused by the action of certain waves upon the eyes and nervesand brain of the person who sees the light. When it is saidthat light is waves, what is really meant is that waves are the physical cause of our sensations of light. But light itself, the thing which seeing people experience and blind people do not,is not supposed by science to form any part of the world that is independent of us and our senses. And very similar remarkswould apply to other kinds of sensations." (Bertrand Russell, Problems of Philosophy, 1912).

2. Bertrand Russell's logical and correct EM wave theory of light did not fit into the new atomic theory. Instead, Einstein's incorrect particle theory not only overcame Huygen's truly inviolable principle of light as EM waves, but would consequently also win over Maxwell's EM wave theory of light, which had been logical for more than 50 years. It would also vanquish Russell's perfect description of the principle of light as EM waves, as well as other sensory experiences all caused by EM waves, which Russell called "sense data". However, this massive evidence of light as EM waves from heated atoms did not match Einstein's new particle theory. Due to Einstein's apparent evidence of the atom, in combination with other impressive insights, the scientific world abandoned the hypothesis of EM waves, which theystill did not know the origin of, for the new hypothesis of atoms with particles in orbit. The atom emitting EM waves was not very well known at the time. An atomic particle theory, on the other hand, was logical because it corresponded to the planets of the universe and because having a particle theory was practical for mathematical calculations. Even in the 21st century the particle remains the main physical theory although all evidence since Huygens in the

17th century clearly proves that the EM wave theory is the principleof the atom.

2.6.24. Einstein (1879–1955 AD) (Light theory: Intromission/Particles)

1. The light was his greatest challenge. Einstein said he had spent all his life trying to understand the nature of light. His interest in the light question arose in the same way as Pythagoras'when he experienced the sunlight reflecting on water in his youngeryears and in the same way as Pythagoras, Einstein came to ponder the mystery of light. Newton had advocated a particle theory of light in 1672, but towards the end of the 19th century everyone knew that light was EM waves; first proven by Huygens (1629–95) and then confirmed by Maxwell (1831–79). Despite Einstein having great respect for Maxwell's theory he still chose to believe that light must be particles coming from the atom, especially when he thought he saw light particles jump out of metal at high temperatures. This was a misunderstanding, though, because what was perceived as light particles was in fact short EM waves that the atoms emit when heated. Even sunlight is EM waves leaving the heated atoms of the sun. The problem of understanding the nature of light at the beginning of the 1900s started a lengthy debate over two different principles of light; as EM waves or as particles[41].

2. The particle theory of light had a massive problem explaining its propagation in air and in space. EM waves have no such problems. For this reason, a particle theory is unrealistic as an explanation for the energy principle of atoms.The particle theory does, however, have a great practical advantage as the "particle language" can be used for mathematical calculations,which cannot easily be done with EM waves. The particle principle created by Einstein and used for atomic energy function has also worked well when it comes to understanding atomic energy mathematics and developing advanced technology. However, forthe understanding of the inner nature and principle of the atom and light, one will find that the particle principle is not relevant.

3. The theory of the atom′s "wave-particle duality" was created as a compromise. The idea was to solve the problemof whether atoms emit EM waves or particles, by instead thinking that they interact. Only EM waves could theoretically explain how light is transported through the vacuum of space. Particles haveno mechanism for this, neither in air nor in vacuum. At the same time, particles were thought to be observed in Einstein's original "light experiment" jumping out of heated metal and this movement was thought to create energy production within the atom. The solution to the dilemma became a compromise in the form of "wave-particle duality". Before the energy content of EM waves could be quantified as easily and practically as "particle energy", it was necessary to maintain the atomic particle theory. Wave-particle duality therefore still applies in the early part of the 21st century as the compromise theory of light, which unfortunately can easily lead to a misconception of the true communication principle of atoms.

4. Towards the end of his life, Einstein concluded thatall the years of pondering the mystery of light had not brought him closer to an explanation. Yes, it had! Einstein's pondering was actually an important part of creating the foundation and development of humanity´s understanding of light. Einstein´s original particle theory, dated 1915, can nowadays be seen as a misconception, but was a necessary, practical and useful theory and development step, since EM waves of light are not easy, if indeed possible, to quantify mathematically. In the end, it is the living creature itself that creates light as an experience in the eyes and brain through EM waves within certain "light frequencies". Light, therefore, is neither particles nor EM waves, but an invention of the brain, just as Pythagoras claimed!

5. Einstein would now realise for sure that EM waves are the communication principle of atoms. It is also Huygens' EM wave theory regarding light, confirmed by Maxwell in 1873, that still applies as the principle of light. Today, a new, important premise must be added: Light does not exist. *Light is nothing but a perception of light when EM waves within a certain frequency range reach the eyes of a living creature.*

2.6.25. De Broglie (1892–1987 AD) & Schrödinger (1887–1961 AD) (Light theory: Intromission/EM waves)

1. De Broglie and Schrödinger advocated EM waves instead of particles when it came to energy and communication principles of atoms.[42] EM waves also better applied to the propagation of light, as light was emitted from highly heated atoms. In the early 1900s, with physicians' greater understanding of atoms, it became increasingly evident that particles could not be the right explanation for the communication and energy principles of atoms. De Broglie and Schrödinger were pioneers for this view.

2. In 1924, de Broglie suggested that what had previously been perceived as photons and electron particles, should now be viewed as EM waves. He meant that EM waves make up the language of atoms and that atoms receive and emit EM waves on a continuous basis. This was demonstrated through experiments in 1927 and was awarded the Nobel Prize in 1929. The EM wave theory was developed by Schrödinger as a specific area of physics called wave mechanics. As the particle theory is the most usable, both mathematically and pedagogically, it has, despite being unfeasible, managed to maintain its hypothetical existence in order to illustrate atomic energy functions in parallel with the EM wave principle. *De Broglie concluded that "wave-particle duality" is an incorrect and misleading compromise, which is only relevant from a mathematical perspective.*

3. CURRENT THEORIES OF LIGHT

3.1.1. Electromagnetic (EM) wave theory or particle theory

3.1.2. There are two official scientific light theories. The electromagnetic (EM) wave theory[43] and the particle theory[44]. Only one of those can be true. The classic explanation is that light is EM waves from a glowing or burning light source, which means that an electromagnetic field spreads through the air from the light source. This EM field from the light source will then be perceived as light by the human eye and brain. The other theory argues that light is in fact particles called photons[45]. Both the "wave theory" and the "particle theory" are seen to be energy carriers and therefore lack matter. It has, however, been impossible for the particle theory to explain how massless particles can exist and how they can be transported through the air.

3.1.3. The particle light theory is a misconception – but a practical one

3.1.4.   The particle theory of light[46] was a direct consequence of the particle theory of atoms. Both were misconceptions. However, the particle theory, mainly advocated by Newton[47] and later Einstein, had the great advantage of being able to mathematically describe the physical world. Even if EM waves was the scientifically correct explanation for causing theexperience of light, the particle theory was the more practical, especially for the quantification of atoms´ energy. The modern quantum-mechanical description was therefore that light in addition to being EM waves, could also be seen as particles called photons. The idea that light could also be considered   particles   is,   however,   a misunderstanding from the end of the 19th century and was initially caused by the observation that light particles seemed to jump out of metal when it was heated.This was, in fact, related to short-term EM waves being emittedby the heated surface atoms of the metal and then incorrectly perceived as individual glowing particles. The particle theory of light does have the practical advantage of being mathematically useful for quantification, but has at the same time made it more difficult to scientifically understand the language of atoms, i.e. EM waves[48].

3.2. Light has no substance

3.2.1. Light is just a perception of EM energy waves as light in the eye and brain. Visible light is a form of electromagnetic radiation, with a wavelength of approximately 390 to 770 nanometres, which is emitted as EM waves from the heated atoms of a light source. In some contexts, the word 'light' is also used for radiation outside the visible part of the electromagnetic spectrum. For light with a wavelength of more than 770 nanometers, the phrase infrared light or IR light[49] is used. Light with a wavelength shorter than 390 nanometres is called ultraviolet radiation or UV light. When the light from a light source is reflected by the matter in our surroundings, we perceive this as colours and shapes.

3.2.2. EM waves that cause visual impressions are often called visible light. This is despite the fact that EM waves arenot visible until they hit the eye. The refraction of light in a prism was first investigated by Newton, who stated that the radiation as such was colourless. The perception of colour can today be explained by the fact that radiation consists of EM waves with different frequencies for each colour. The transitions between different frequencies happen gradually, but the spectrum is traditionally divided into seven colours. Usually, the radiation, whichhits the eye and causes colour perception, is a mixture of many different wavelengths.

4. THE EXPLANATION OF LIGHT

4.1. Light is an invention of life

4.1.1. Without the eyes and brain of a living creature, there is no light. Light does not exist as a separate phenomenon,as such. In order for light to be perceived, something does not only need to burn or glow, but the eyes and brain of a living creature are also required in order to transform the emitted EM waves of the highly heated atoms into an experience of light inside the brain. All sensory experiences are actually created as experiences inside the brain, when EM waves from the atoms of the outside world reach each individual sensory function – eyes, ears, nose, touch, taste. Light therefore does not exist as a physicalphenomenon on its own, but is an invention of life. Instead,light is EM waves from a light source with a high energy frequency emitted by burning or glowing matter, which a living creaturewith eyes and brain will experience as light.

4.1.2. Being enlightened is an expression for knowledge. At the same time, the ironic thing is that man does not knowwhat light is, but the lack of knowledge when it comes to lighthas so far not been perceived as a problem. Certainly, man hasnot been able to physically define light, but questions like the Big Bang and the nature of light may have been assumed to belong to a category that simply cannot be explained. Humans have always been curious about how everything came about and how it works and as they studied light, they discovered things that didn´t really tally and which suggested that something is not quite right. Particularly, the reflection of light on water gave reason for inquiry.

4.2. The scientific history of light sincethe 17th century

4.2.1 During the 1600s, light could be explained as EM waves. This became possible through a theory of Huygens' which was subsequently proven from a physical aspect during the 1800s and explained by Maxwell, but still without understanding how or why EM waves are created and exist naturally.

4.2.2. At the beginning of the 18th century, Newton arguedthat light is, in fact, particles. Due to Newton's strong statusas a prominent mathematician and scientist and in combination with the simpler, more practical comprehensibility of the particle theory, it was only natural that Newton's particle theory of light became more accepted than Huygens' EM-wave theory.

4.2.3. Towards the end of the 19th century, EM waves were developed into wireless radio transmissions. As the experience of sound is created when EM waves from atoms exposed to physical impact reach the ear, the experience of light is created when EM waves from highly heated atoms reach the eye. The principle and "technology" for sound based on EM waveswere developed mainly by Maxwell (1873), Hertz (1888), Tesla (1893) and Marconi (1896) and led to the use of EM energy waves for wireless radio transmissions[50]. However, as Einstein's particle theory became more intelligible when it came to describing the energy concept of the atom, it came to take over, quite erroneously, the already logical and proven understandingof EM waves as the language of atoms. As it was easier to understand the particle theory, this meant that all of the already established and convincing evidence of EM waves as the languages of atoms was abandoned. Since then, science has triedto solve the problem by considering both theories as valid and applicable, which may appear as a practical solution, but as a matter of fact is untrue because it contradicts facts and end up creating confusion in the scientific field. The particle theory is indeed necessary for mathematical quantification, but is not a correct description of reality, which is undoubtedly the EM energy wave. Particles simply do not exist in the atom.

4.2.4. In the beginning of the 20th century, Einstein developed an atomic particle theory. Newton's mathematical and practical particle theory of light became a model for Einstein's theory development relating to atomic energy production and energy propagation. However, this was at the expense of the real physical principle of light, i.e. EM waves, which now was abandoned even though Maxwell had proved this extremely convincingly towards the end of the 19th century. Maxwell's correctEM wave principle did not, however, help as a mathematical understanding of the atom, because there was no way of quantifyingEM waves.

4.2.5. Since the 1930s, atoms´ energy function has beenviewed both as particles and EM waves. This was the right solution for mathematical understanding of the atom, but at the expense of true understanding of the energy language between atoms, as well as the universal communication principleof nature, which is solely EM waves. Since 1934, both the particle theory and the EM wave theory of atoms have been accepted in parallel, but without any scientific clarity regarding which oneis actually correct and why there are two different approved theories of light. It is, in fact, only the EM wave theory that is scientifically valid and applicable. A particle theory cannot explain the physical principle of light, while the EM wave theory can. On the other hand, the particle theory must be used for mathematical understanding of light.

4.2.6. At the beginning of the 21st century, the true existence of atoms´ EM waves are still more or less ignoredin favour of the particle theory. Nor is it generally understoodthat EM waves are emitted by atoms. The importance of

EM waves for all kinds of communication in nature is in fact just as fundamental as they have become in the technical world of humans. The reason why natural EM waves have not been fully recognised is that all practical physics can most easily be mathematically described with "particle physics". Despite the fact that EM waves are the true energy and communication principle, the atomic particle theory has become the dominant energy concept because of its more practical, mathematical understanding. For this reason, man has not truly realised that EM waves are, in fact, the natural energy principle of atoms and the communication principle of everything.

4.3. The experience of light is caused by EM waves

4.3.1. Light particles do not exist. The atomic particle theory was classed as valid at the beginning of the 20th century. The particle theory for atoms was mainly developed by Newton in the latter part of the 17th century, but by the end of the 19th century had been proven wrong by the EM wave theory, only to be revived again by Einstein in the early 20th century. This was not due to a particle theory of light unquestionably being correct, but simply that a particle theory was mathematically more practical and useful than EM waves. The rumour of Einstein's intelligence as a scientist also spread rapidly and the atomic particle theory became his most famous proposal. It was based on the discovery that a heated piece of metal emits a spark at a certain temperature. At least,that is what it seemed like to the human eye, which was the reason behind the atomic particle theory, but which actually was a short EM wave. That heated, glowing particles inside an atom, with a high speed of movement, could jump out of matter was not an illogical assumption as such, but it was incorrect. The fact that the particle theory was incorrect can today be demonstrated mainly by the operating principles of the eye and vision. What the eyes receive and transform through eyesight and the brain into a perception of light, is in fact EM energy waves emitted by highly heated atoms, e.g. from the sun.

4.3.2 Einstein became world-famous in 1919 by correctly predicting a solar eclipse. This was due to one of the most spectacular scientific experiments, or rather demonstrations,in history. Einstein was the scientist who came most of all to symbolize scientific development in the first half of the 20th century. Like previous prominent philosophers, he had special skills in mathematics and physics. As the second promoter of the atomic particle theory – the first being Newton – Einstein gaineda unique position in the early 1900s as the most prominent of physical scientists. His first major success occurred in 1919. Priorto a solar eclipse northeast of South America, he had predicted that it would be possible to prove that both light and matter consist of particles, as gravity would cause the light from a distant star to be diverted towards the sun. With the whole world´s attention,this actually did happen, which immediately made Einstein world-famous and he was celebrated accordingly with a tributeparade on 5th Avenue in New York. This commotion and excitement could be considered apt, because the particle theory of light thus appeared to be proven. With this, the question of whether atoms give off particles or EM waves no longer seemed a matter for discussion. Einstein, with his particle theory, simply emergedas the great winner. However, what was assumed to be proven at the solar eclipse, that both light and mass are particles, was in fact ahasty conclusion. Today it is understood that everything that exists,

i.e. atoms, is actually all energy and the reason for the light from the stars being drawn towards the sun was actually that a larger energy attracts a smaller.

4.4. The explanation of light as EM energy wavesis proven since 1929

4.4.1 The discovery by Huygens and Maxwell, of light being EM energy waves, finally disproved the particle theory. Einstein´s credibility and fame from the solar eclipse demonstration in 1919 was seemingly impressive, but could not disprove the solid evidence supporting the EM wave theory in 1929.EM wave and particle theories of the atom have since been consideredequal, i.e. both theories are applicable. It is, in fact, only the EM wave theory that is physically relevant, as particles do not exist. Since particles don´t exist in atoms, the particle theory is utilized only for its practical comprehensibility and mathematical description. If a "quantification" of EM waves and their energy could be developed for mathematical purpose, the basically unrealistic "particle physics"would no longer be needed. The particle theory will actually alwaysbe needed in parallel, in order to mathematically illustrate the scientifically correct EM wave theory. The two theories are doomedto cooperate in parallel.

The term "wave-particle duality" is a misleading concept. It indicates that both theories are valid as an atomic energy principle, which is not the case. When the theory of quantumphysics was created in the early part of the 20th century, light was given a particle explanation, partly due to the illusion of jumping particles from heated metal, which actually instead was short-term EM waves leaving the surface atoms. It was then considered logical to assume these were energy-creating particles from within the atom which, if heated, would leave the atom. That atoms would provide energy in this way was considered logical, but at this point in time, no one could imagine the immense built-in force that was later demonstrated by the atomic bomb. Such vast amount of energy being able to form through particles in orbits within atoms must instead be regarded as unfeasible. As an alternative to the explanation of light through the atomic particle theory, theories based on EM waves were developed. The most convincing was formulated by de Broglie and Schrödinger, rendering the Nobel Prize in Physics to both of them in 1929. Despite the EM wave theory being the correct energy principle of the atom, the system inside the atom is still described in the traditional way, with its energy being created by particles in tremendous revolution. The first question about this principle would be how these "tiny-tiny" energy-factories can create energies with the power of atomic bombs. There is presently no official theory how the energy in each atom is being created, although a system where cosmic EM energy waves continuously load energy into all atoms seems less improbable than the present particle theory. It is important to emphasize, though, that atoms' communication principle is

EM waves.

4.5. The speed of light

4.5.1. From Pythagoras´ light theory to Aristotle´s light theory and to current light theory. Pythagoras, in approximately 500 BCE, understood through observations and mathematical theorems that no physical light could exist in the air and that the light must be a perception of light inside the human eye and brain. Aristotle, however, presumed approximately 200 years later that the world exists in the way that human senses perceive and experience it. That was principally the accepted theory for a long time, right up to the 16th century AD, when Galileo was the first man to attempt an experiment measuring the speed of light. But even the modern understanding of light is more or less basedon Aristotle's simple and seemingly natural, theory. There wasin fairness no other alternative, as Pythagoras' original theory was seen as unfeasible and was therefore "adjusted" in various ways, but remained implausible. As the true nature of light was a difficult problem to solve, the science of light has ended up being more about how light creates human vision and not what light really is. The principle of light has, therefore, remained Aristotle's light theory dating back approx 2,300 years, namely that light is simply light. In modern times, scientists have described light bothas EM waves within certain wavelength ranges, which humans perceive as light and also as particles called photons. The latterhas been useful when it comes to quantifying light, but not when it comes to explaining what light really is.

4.5.2. Light does not have a continuous speed – just an initial "phase velocity". EM waves of light from a light source only have an initial propagation speed, called "phase velocity".This means that EM waves of light reach a certain distance fromits source, depending on the energy of the light source and then remain constant. Experiments carried out to measure the speed of light only refer to this phase velocity of EM waves. What is known as the "speed of light" is therefore the initial velocity of EM waves from the light source before they either reach the maximum length, or become stopped and stationary. For example, when EM waves from the sun reach planet Earth, it is no longer a matter of speed, because the waves are now temporarily stopped. The speed of light in a vacuum has been considered the fastest possible speed. As will be explained below, this is not actually true, because it has been revealed that light in certain materials propagates faster than in a vacuum.

4.5.3. Since 1983, the speed (propagation) of light has been officially defined as 299,792,458 m/s. The speed of light in a vacuum is falsely proclaimed the fastest possible speed and Einstein's relativity theory states that nothing can be faster than the speedof light in a vacuum. It has also been presumed that the speed of light in a transparent medium, such as glass, water and air, is slowerthan the speed of light in a vacuum. In another information source, the speed of light in water is stated as 25 percent faster than in vacuum, while yet another quote says that the speed of lightin water and in a vacuum is the same speed and yet another thatlight is slower in a vacuum than in water, i.e. that light is faster in water than in a vacuum! The "speed of light" issue is obviouslya confusing one. However, the speed of light in a vacuum (c) is internationally accepted as a physical constant and is 299,792,458 m/s. The speed of light is considered independent of the observer's movement - two different observers will always measure the same speed, regardless of how they move in relation to each other. "The speed of light in a vacuum" is, although with great uncertainty, proclaimed the fastest possible speed.

4.5.4. The light from the moon, sun and other stars is – once established – distributed with no delay. Since the light distribution from the stars as

EM waves, after being established, has a permanent "non-motion" establishment in space, its EM waves constitute a permanent direct transfer to planet Earth and thus appears without time lapse. This means that when a person on Earth looks at the starry sky at night, the star's elongated EM waves are constantly and immediately accessible to the eye and offer an instant light experience for the observer. The sun is also a star, so this principle applies to sunlight as well. When the sun is on the other side of Earth, its light is reflected from the moon and creates moonlight for the observers on the dark side of Earth. The elongated EM waves emitted by the stars are always permanently established in space, with extension as far as its supporting energy can allow. The light from the stars, including the sun, has therefore normally no speed and is stationary, except when starlight is being extended again after having been temporarily obstructed. Stars' permanently stretched-out EM energy waves in space reach planet Earth and enable man to see the light source in return. Once the propagation of EM waves from a light source has reached its stop, the communication of its energy to the stop is

In contrast, man-made electronically transmitted sound from the moon to the Earth takes approximately 6 minutes. This is because the EM sound waves utilized are created and transmitted based on "man-made technology", while the light transmission from stars is supported by the immense powersources of the burning stars. During the day, the distant stars far away from Earth are not visible because the sun's EM waves oflight dominate the atmosphere around Earth. The sun´s EM waves are permanently extended and therefore have no speed and are just like the EM waves of all other stars – mainly permanently elongated and established in space and only hindered due to temporary blocking by other stars and planets. When, for example, the maximum extension of sunlight waves is established, its EM waves are an integral part of the solar energy system and no longer have any movement, but are considered static EM waves from the sun.

4.6. Observations that challenge the understanding of light

4.6.1. Light can be created in other ways than through fireand glow. The scientific research of light is called "optics". Certain chemicals can produce light called "chemo-luminescence". It is known that light can radiate from certain animal species, e.g. fireflies, a process called "bioluminescence". There are in fact many chemicalsubstances and natural preparations that create what man perceives as light. In a suitable medium, it is also possible to create an EM wave from a radioactive source that moves faster than the speed of light.

4.6.2. The EM waves of light create a physical force on matter. This was previously explained by the particle theory of light, with physical particles hitting another physical object. Light is actually always radiating from something physical, not as physical particles as such, but as energy in the form of EM waves. The explanation for this is that even if light does not contain as much energy as matter, its energy is enough to cause an impact on whatit hits. Such a light impact is, for instance, demonstrated by the physical impact of sunlight on human skin and the bleaching of colours exposed to light.

4.6.3. "Cherenkov light"[51] moves faster than light. It has actually been considered possible for particles or EM waves tomove faster than light inside a medium. When this happens, aform of radiation is emitted, the so-called Cherenkov radiation. Cherenkov radiation is a phenomenon that occurs when a charged particle in water moves faster than the speed of light. The speed of light in different transparent media is usually slower than ina vacuum. The phenomenon has been named after the Russian physician Pavel Cherenkov, who discovered it in 1934 and was awarded the Nobel Prize in Physics in 1958. Such a radiation had actually been theoretically foreseen and published by English scientist Oliver Heaviside in 1889.

4.6.4. Light only has an initial extension velocity until it reaches its longest spread.

The speed of light is, in fact, the velocity of light from its initial burning or glowing phase and the light's EM waves are to be regarded as the outstretched energy extension system of the light source's atoms. The light is thus almostimmediate and then static. The challenge with the Cherenkov radiation was and still is, that its propagation in water or other medium is faster than the velocity of light in a vacuum, which, according to Einstein and his relativity theory, was assumed to be the fastest possible speed.

4.6.5. According to Einstein, the speed of light in vacuum is the fastest possible.

The discovery and understanding of so-called Cherenkov radiation have meant that light can in factmove through matter at a faster rate than in a vacuum, which contradicts the relativity theory. This insight came in 1934, when it was discovered that atomic particles in matter could move at a greater speed than the speed of light and thereby produce visible Cherenkov radiation. In 1958, Cherenkov was awarded the Nobel Prize in Physics for this discovery. One issue with the understandingof Einstein establishing the official speed of light at 299,792,458 m/sis that this does not refer to a continuous speed of light, but only to the initial "phase velocity" of light. When light has reached its maximum propagation, its EM waves no longer have a speed and arestatic EM waves stretching out from the light source.

4.6.6. The Cherenkov light radiation contradicts Einstein's theory of relativity. The challenge of Cherenkov radiation is that its speed in water, or in another medium, is actually faster than the speed of light in a vacuum, which according to Einstein had been supposed to be the fastest possible speed.

5. ATOMS ARE THE ENERGY SOURCE OF LIGHT

5.1. The atom

5.1.1. The Greeks, 6th – 5th century BCE, gave the atom its name. It stems from the Greek word "atomos" and means "indivisible". The earliest documented theory about the tiniest unit of matter was developed by Leucippus and Democritus during the heyday of Greek philosophy, in the middle of the 5th century BCE, which was then kept alive and advocated by Epicurus (341–270 BCE). They meant that a pieceof solid matter theoretically can be split in half, in smaller and smaller segments, until you end up with the smallest possible unit, i.e. the atom, which cannot be split. After the era of Greek philosophers, from approximately 3rd century BCE and onwards, these atomic theories were forgotten just like much of the rest of scientific philosophy, at the same time as religions and their teachingswere growing more influential. It was not until the 17th century, almost 2000 years later, that science was resurrected as a topic. The original theory of the atom being indivisible was now proven correct. Everything in nature can be split in half until you reachthe smallest possible unit which is in fact indivisible. Today it has been confirmed that atoms can only be split and destroyed with an explosion, a blast, which furthermore reveals the incredible energy stored in the atom.

5.1.2. There are 118 different types of atoms that can be combined and in an infinite number of ways. They makeup all the different substances that exist, which humans perceive as matter, gases and liquids. Atoms are extremely small. Ten million atoms in a row would still only measure 1mm in length, which means that an average atom has a diameter of about 0.0000001 mm, i.e. one ten-millionth of a millimetre. Even the air that surrounds planet Earth is made up of atoms, whereof 99 percent are nitrogen and oxygen atoms, which makes it possible for living creatures to breathe.The stars and planets in the universe, including planet Earth and its air, all consist of atoms. Historically, atoms have only been regarded as "nature's building blocks", because their communicative role was not yet known. Outside Earth's atmosphere there is a vacuum, which could in fact be called empty space in the absence of atoms, but is nevertheless full of emitted energy waves coming from the atoms of the stars that allow man upon Earth to see the stars in return.

5.1.3. The atom is energy – but what is energy?
Energy is the most fundamental term in physics and represents the energy that atoms contain and that mankind and everything in nature use to enable life on Earth through the basic needs of food, light and heat. Energy has the quality, at least on planet Earth, of being renewable and can therefore alternate between different usages such as heat, food or motion energy. To a great extent, planet Earth's natural movement and climate contribute to the creation of energy and also its continual transformation from one type of energy to another. The energy that builds everything on Earth is, in fact, to be seenas two completely separate types of energy – partly what is naturally available for the normal needs of mankind, but also, as man has learnt to understand and utilise over the last century, to open up an energy supply that can be seen as eternal. This is

about the built-in energy of certain atoms, i.e. atomic energy; an asset of indefinite size for human society, but at the same time, an operational risk and a risk of war due to its potential use for destructive purposes.

5.1.4. Atoms consist of energy – not particles of mass. At the beginning of the 20th century, Einstein's particle theory became the accepted explanation of atoms and light. Since Einstein's particle theory proved to be more understandable whenit came to describing the energy concept of the atom, it started to incorrectly override the already logical and proven understandingof EM energy waves being the language of atoms. Since then, science has tried to solve the problem by considering both theories as valid and applicable, which may appear as a practical solution, but is in fact untrue because it contradicts facts and ends up creating more confusion in the scientific field. The particle theory of atomic energy communication is admittedly useful for illustration and for mathematical quantification, but is not a correct description of reality - which is atoms extended with EM waves.

5.1.5. Einstein & Eddington [52] unwittingly proved the particle theory to be incorrect. This happened in 1919, through one of the world's most important and most spectacular scientific experiments, in connection with a solar eclipse. The intention was to prove that gravity is caused by the mass of matter. It also turned out that the expected gravity and the diversion of the light of a star towards the sun, at the time hidden behind the moon, really did occur. This was seen as proof that both starlight and the sun must consist of particles with mass. Today it is known, however, that as light is energy in the form of EM waves, the starlight could not have been diverted towards the sun due to gravity caused by the attraction between masses of real gravity. Instead, gravity must be due to the fact that both light and matter of the sun consist of energy.Only vacuum has no energy. The old theory that two masses of matter exert mutual attraction in accordance with the law of gravity must therefore be corrected to gravity being caused by the energy content of matter, i.e. atomic energy. This experiment can therefore be considered as evidence that everything that exists is, in some way,only atoms with different degrees of energy.

5.1.6. Nothing has a real weight in the universe. Atoms have no weight. Everything is weightless energy. The weight, gravitationor force of gravity are, in fact, an attraction force caused by the difference between the energy content of an object and the energy content of the planet or the star on which the object is located. This has, for instance, been proved by astronauts' visits to the Moon. In fact, already in 1924 it was quite clear that planet Earth and the entire universe do not consist of atoms of mass with its own weight,but of atoms that are nothing but energy. This conclusion waspresented by de Broglie, for which he was awarded the Nobel Prize in 1929. However, the idea of matter with mass and weight lived on, as this is how man naturally perceives the world. It might seem strange that de Broglie's qualified and recognised insight into this topic failed to change the ancient idea of atomic construction, from mass of real weight to nothing but energy, despite the fact that mass was proven not to exist. Attempts have been made to find some mass inside the atom, but still by year 2017, no mass has been found. The search has nevertheless continued and scientists are still looking for the mass believed to exist inside the atom, because everything falls to the ground. What is to this day called "gravity" is actually not linked to the supposed real weight of matter, but is an effect of planet Earth's energy in relation to the energy of a loose object. A larger energy exerts attraction on a smaller energy.

5.1.7. For a sum corresponding to 6.5 billion USD,the "Large Hadron Collider" [53] was built in CERN, Switzerland, 2008, with the aim of establishing whether there is any

mass of real weight inside the atom. No mass was found. The scientists found that the so-called weightof matter seems to build on an entirely different principle than mass with weight; namely that a larger energy, such as a planet, attracts a lesser energy, like living creatures or loose objects.It has also been known since 1785 that "Coulomb's law"[54],a mathematical formula for the force of attraction of different energy units, is actually identical to "Newton's law", which is an equation for gravitation. Newton's law is, however, no evidence that a physical mass causes gravity, which Coulomb's law is when itcomes to energy as a gravitational force.

5.2. EM energy waves of atoms is the communication system of the universe

5.2.1. Atomic EM waves communicate the interaction **between everything that exists.** This fundamental feature of the universe was detected by Maxwell, but only in terms of EM waves' existence and usability for remote communication on planet Earth. Maxwell concluded that EM waves must also be the principle of light. All communication in nature, as well as between living creatures, is conveyed via EM waves. For example, light from thesun or other light source flows into the eyes, sound from every sound source flows into the ears of a living creature, the taste of something edible fills the mouth, the experience of touch, the perception of a scent, etc. Huygens had already established that light was EM waves,but at the time this theory had not been utilized as an explanationfor other sensory experiences. The transmission of sound, taste, touch and smell to each individual receiving organ did not, in fact, have any explanatory theories at all. Touch had, for example, nothing but the practical explanation of physical contact, much like the physical transmission of sound, taste, sight and smell to the receiving body. The problem or challenge was to be able to give a detailed description of the principle behind the physical transmissionof primarily light, sound and smell that takes place in the air, while taste and touch could already, quite logically, be understoodas direct physical contact.

5.2.2. Maxwell discovered atomic EM energy waves in 1864. He arrived at the conclusion that atoms communicate through EM waves and that EM waves of sound should adhere to the same principle as light. The principle relating to EM energy waves of sound was then developed by Maxwell, Hertz, Marconi and Tesla and later applied in wireless radio transmission from 1896 onward. At that time, scientists did not know how EM waves occur naturally,which is still fairly true to this day. The latter may seem strange, because EM waves handle all communication between atoms in the universe, i.e. all communication including light, scent, taste and sound, as well as all kinds of physical transmissions or movements. The sound of voices from all living creatures, as well as sounds fromevery sound source, are in fact also created and conveyed in the air asEM energy waves, but are not perceived as sound until they reach the sense of hearing of a living creature. If a tree falls in the woods and no living creature is around, there is no noise because there isno living creature with ears and a sense of hearing that can pick up on the EM waves that occur with the fall and transform these wavesinto an experience of sound in the brain. One of the issues that have been particularly hard to explain is how light, sound and scentsare physically transported in the air, while taste and touch have beenunderstood as direct physical contact. All sensory experiences, including direct contact such as taste and touch, are communicated and, in fact, caused by a physical transfer of energy through EM energy waves. For such transmission of "sense data" within defined frequency ranges, there are recipient organs of a living creaturethat convey the data to a corresponding experience inside the brain.

5.2.3. **At the beginning of the 20th century,**

there was a need for a mathematically calculable particle theory of atoms to complement the EM wave theory of light. The problem with the EM wave theory was that it is impossible to quantify mathematically. The particle theory applied primarily by Einstein, on the other hand, was a practical method for atomic energy calculation, but to the detriment of the true principle of light, which is EM waves. However, with the correct understanding of atoms, that their energy exchange is done through EM waves, a mathematical understanding of the atom cannot be achieved because EM waves cannot be quantified like particles can. Boththe EM wave theory and the atomic particle theory have therefore been accepted as valid and applicable, even though it is only theEM wave theory that is scientifically correct. The particle theory was,in fact, a necessary measure for mathematical reasons. A particle theory cannot explain light, though; only the EM wave theory can and is therefore quite evidently the only true energy theory.

5.2.4. EM energy waves are the communication principle of atoms. This undeniable communication principle has, however, been mainly ignored as the true interpretation and understanding of nature's interactions, because for mathematical reasons, a particle theory was preferred to explain the atom. Although this has been a choice from a practical standpoint, it has meant that the scientific logic and understanding of the principle behind atomic energy communication have become neglected and incorrect. All atoms, i.e. everything that exists, both receive and emit EM waves in nature's interactions. All kinds of communications received by the senses of living species are based on nature's system of EM waves emitted by atoms, which when reaching the sense organs of living species cause the physical impression of sound, sight, touch, smell and taste. Notonly do atoms build everything that exists, they also constantlyemit energy radiation in the form of EM energy waves with different frequencies. Especially with the principle of the ever-communicating EM waves of atoms and how information from the outside world flows to the minds of living beings, the EM wave theory should be regarded as a main fact, not a theory. Light gets reflected via matter and when the EM waves of thislight reach the eye, images are created in the brain depicting the reflection surfaces of the person's surroundings. In order for light and colours to exist, a living creature with eyes and brain is required. The brains and senses of living species, which are different for each species, interpret the world and create a worldview for that particular species. Where there is no living creature with eyes and brain, no light or colours actually exist. The principle and "technology" for sound through EM waves were primarily understood and developed by Maxwell, Hertz, Marconi and Tesla and led to the use of EM energy waves for wireless radio transmission from 1896 onwards.

5.2.5. Einstein's particle theory was much easier to grasp than the EM energy wave concept of atoms. Which is why the particle theory came to quite wrongly overshadow all previous evidence and understanding of atomic communication as EM waves. This was however not due to the particle theory being the correct theory, which it is not, but that the particle theory was bettersuited to a mathematical understanding. As a theory of atomic energy production, it lacks an explanation of what would set the particles in motion. As Einstein's particle theory appeared to be more comprehensible when it came to describing the energy principleof the atom, it actually replaced EM waves as the language of atoms in the 1920s. The better mathematical understanding that the particle theory brought meant that all the established and convincing evidence of EM waves as the languages of atoms was abandoned. Since the 1930s, science has tried to correct this by accepting both theories as valid, which may appear to be a practical solution but is, in fact, not practical at all as they contradict each other, which in turn creates confusion in the scientific subject matter. Although the particle theory of atomic energy communication is necessary in order to illustrate and quantify mathematically, it is not a correct description of reality, which is that our world is made up of atoms with EM energy waves.

5.2.6. The principle of light as EM waves from highly heated atoms is today proven correct. The historical description of Pythagoras' theory, i.e. that light is a perception inside the brain, which was the main light theory among philosophers for several centuries, can today be seen as true. First of all, it has been established that light is not something that exists physically in the air, but is EM energy waves from a light source. It is therefore perfectly correct to say that the human eyes and brain create the light, or rather the experience of light. Where there is no living creature with eyes, no light exists either. Light does, in fact, never exist as light in the air, but instead it is each individual that creates an experience or perception of light with its eyes and brain by receiving EM energy waves from highly heated atoms. A decisive proof of this is the light of the rising or setting sun reflecting on water, as this is always perceived by each observer as a glitter path seemingly heading exclusively to each individual.

necessary in order to illustrate and quantify mathematically, it is nota correct description of reality, which is that our world is made up of atoms with EM energy waves.

5.3. The understanding of light opens the doorto a new world view

5.3.1. Is light EM waves or particles? The answer is – neither. The scientific understanding is that light can be viewed as both EM waves and particles, but this is for practical reasons as it has been uncertain which elemental form light actually has. When people talk about light as EM waves, the general understanding is that EM waves are light, which is not the case. In fact, light isjust perceived as light when the EM waves representing light reach the human eye and brain. The EM waves of light, from the sun or any other source, are therefore never real light but only EM energy waves perceived or experienced as light when they hit the eyesand brain of a living creature.

5.3.2. Today, in the 21st century, it is understood that lightis not real light. To put it simply, light never exists as real lightin the air. In fact, light is never anything but a perception of light inside the human eyes and brain when EM energy waves fromhighly heated atoms of a light source reach one's eyes. The other sensory experiences, including sound, taste, scent and touch, also never exist as real physical phenomena, but follow the same principle as light, i.e. in the form of EM waves emitted by a source and received by each individual sensory organ, while touch is a combination of EM waves and a physical sense. For each transmission of "sense data" in the form of EM waves, coming from the atoms of the surrounding world and entering into a living creature, there is a receiving sensory organ that transmitsthe information to the brain where the conscious experience is created. Each living creature with eyes is therefore creating its own light when the eyes receive EM waves from a light source, whichis not really true light as such, but something the brain transforms into a perception of light. Similarly, the other sensory experiences are conveyed and received as EM waves from the atoms of the outside world to each respective organ, and from there to the brain.

5.3.3. The different sense- and brain functions of the living species create an individual worldview for eachspecies. The fundamental and all-encompassing language of nature is energy in the form of EM waves communicated from the atoms of everything around us to the senses of different living creatures. EM waves/radiation are reflected from everything that is radiated by light energy, and when the reflected energy waves reach an eye, they create colour images inside the brain according to the reflection surface. Each shade has a certain energy frequency that the observer's sense of vision interprets as colour. Therefore, in order for light and colours to exist and be experienced, a living creature with eyes and brain is required. If no living creature exists that can transform the solar EM waves into light and image with its eyes and brain, there will be no light. If a tree falls in the woods and nobody is there, there is no sound.

5.3.4. The way eyesight works by the eyes receiving EM waves proves that a particle theory of light is not correct. The particle theory was based on a misinterpretation of the earliest experiment of quantum physics, carried out by Planck and Einstein, when metal was heated to such a point it looked as though it emitteda particle which was, in fact, a spark in the form of a short-termEM wave. That heated, glowing particles inside an atom would create light through high rotational speed was not actually an illogical assumption at the time, but was incorrect. What the eyes receive through the sense of vision and transform into an experience of light, is actually EM waves of light being reflected from the atoms of everything around us.

6. EM WAVES FROM ATOMS CREATE ALL 5 SENSORY EXPERIENCES

	PAGE
6.1. Light and all other sensory experiences are caused by EM waves	114
6.2. In 2014, scientific evidence proves sound is EM energy waves	121
6.3. Light, sound, colour, scent, taste and touch are all conveyed by EM energy waves	125

6.1. Light and all other sensory experiences arecaused by EM waves

6.1.1. Light and eyesight are sensory experiences based on EM waves – sound, taste, touch, and smell are all based onthe same principle. All sensory experiences are created and conveyed by atomic EM waves, either as direct physical contact or through the air. The experience of taste, for instance, takes placein the gustatory system, when EM waves are being emitted from the atoms of the food as it is processed in the mouth, while the experience of a scent is created in the olfactory system by EM waves being emitted into the air from odour-creating matter, and the sense of touch by EM waves at the point of physical contact. Atomic EM energy waves have a much greater role than previously assumed. In fact, EM waves have become almost completely ignored, because the particle theory of physics was incorrectly used as the explanatory model of the atom.

6.1.2. Everything that exists in a physical sense consists of atoms that emit EM waves. This principle applies to all senses of all living creatures as they receive and interpret EM waves. Sound, for instance, is not a real sound in the air, but always justEM waves that the human brain and sense of hearing perceive as sound. The most obvious proof of EM waves being the language of atoms and energy carriers, is the EM waves from the sun's extremely heated atoms, which are commonly but falsely called rays of light or sunlight, and which when they hit planet Earth are perceivedas light by the human eye and brain. The understanding that atom's EM waves are nature's way to communicate and are the language of everything, was for a long time hindered by the particletheory advocated by physical science, as particles were the most understandable for quantity assessment description or such like.The particle theory, however, is not feasible for the simple reason that there is no valid principle for the particles' movement through the air, while EM waves are the perfect system.

6.1.3. The particle theory of light was a consequence of the particle theory of atoms. Both were misunderstandings, especially due to the lack of conveyance principle in the air for the presumed particles. Although EM waves were the most convincing explanationfor light, the particle theory of light was the more practical for the quantifying of atomic energy. Modern quantum mechanics[55], therefore, described that light, apart from being EM waves, also could be considered particles called photons. The idea that light could be considered particles, though, is another misunderstanding dating back to the late 1800s, initially caused by the observation of light particles seemingly jumping out of a piece of heated metal. This was, in fact, short term EM waves which when heated, were emitted by the superficial atoms of the metal and were then perceived as individual glowing particles. The particle theoryof light does, however, have the decisive practical advantage of being mathematically useful for quantification, but has at the same time interfered with true scientific understanding of the atomic language, which is nothing but EM waves.

6.1.4. Einstein's particle theory was much easier to grasp than the energy concept of atoms. Which is why the particle theory came to quite wrongly overshadow all previous evidence and understanding of atomic communication as EM waves. As a theory of atomic energy production, it lacks an explanation of what would set the particles in motion. As Einstein's particle theory appeared to be more comprehensible when it came to describingthe energy principle of the atom, it actually replaced EM wavesas the language of atoms in the 1920s. The better mathematical understanding that the particle theory brought meant that all the established and convincing evidence of EM waves as the language of atoms was abandoned. Since the 1930s, science has tried to correct this by accepting both theories as valid, which may appear to be a practical solution but is, in fact, not practical at all asthey contradict each other, which in turn creates confusion in the scientific subject matter. Although the particle theory of atomic energy communication is necessary in order to illustrate and quantify mathematically, it is not a correct description of reality, which is that our world is made up of atoms sending and receiving EM energy waves.

6.1.5. Sound was the first piece of evidence that EM waves are the conveyor of the human sensory experience. Huygens had this insight in 1678, which was later confirmed and realisedby Maxwell in 1861, and resulted in a communication revolutionon Earth via radio and telephone. Sound, which is created insidethe sense of hearing by EM waves from vocal cords or other sourcesof sound, could now be conveyed as EM waves from a radio stationthrough the air, end up in a destination of your choice anywhere around the globe, and then transformed back to the original sound of a voice inside the listener's auditory organs. This revolutionary principle has been seen as more of a technological discovery than a discovery of the natural principle of EM waves that it is really based on. With the help of Maxwell, the world of science then accepted the fact that light is EM waves with a certain wavelength. Why EM waves occur naturally and wherethey come from, however, were unknown and are, in principle,still unexplained 150 years later, because physical sciences wronglyaccepted an atomic particle theory instead of an EM wave theory. However, the main clue to the insight about EM wavesis light. The understanding of the principle of light and vision also provides an overall clue to the other sensory experiences - touch, taste, smell, sight and sound.

6.1.6. As sound and light are sensory experiences based on EM waves, the other sensory experiences are most probably based on the same principle. Atoms of the speech organ do not emit real sound as such, but EM waves that are transformed into an audio or voice experience inside the sense of hearing and brain of both "transmitter" and receiver. Neither incoming soundto one's own hearing organ, nor outgoing voices from the atoms of the vocal cords, are real sounds; only silent EM waves. Sound neverexists as sound in the air, but merely as EM energy waves, which are perceived as sound in the ears and brain of a living recipient. The EM waves of sound are created in the vocal cords of a human, for instance, or in musical instruments and through punches, bumps, impact, vibration, air flow, or friction. In every situation where atomsare exposed to physical impact, EM waves are created by and emitted from the affected atoms and propagate through the air or matter, andare then perceived as various sounds when they reach the hearing organ of a living creature. What is true for all sound is that it is neverreal sound as such, but only silent EM energy waves created inside the sound source, perceived as sound when they reach the hearing organ of a living creature. What living creatures experience as taste, smell or touch is conveyed in a similar way. The taste experience, for instance, takes place in the gustatory sense inside the mouth, with EM waves being emitted from atoms ingested and processed inside the mouth.

6.1.7. Einstein's particle theory on atomic energy capacity would, at the start of the 20th century and in conjunction with the discovery of the atom, be seen as the accepted explanation. Einstein's theory would more or less overshadow or even eliminate all previous evidence and understanding of atomic communication being EM waves. This was actually not because the particle theory was right, which it is not, but because the particle theory was more comprehensible and more suitable as a theoretical and mathematical explanation. As a theory of atomic energy production it does, however, lack an explanation of what wouldset the particles in motion.

6.1.8. The EM energy waves of the sun are reflected by all superficial atoms. Everything that exists in the human world both emits and receives EM waves as a part of nature's interactions.This function creates nature's total, continuous, physical systemof cause and effect. Everything on planet Earth consists of atoms that exist in 118 different types of energy, which in turn - in various combinations – make up everything in the human world. Not only do atoms build everything that exists, but their emitted EM waves arealso used by the senses of living creatures in order to experience sight, smell, taste, hearing and touch, which makes up the ability of living creatures to experience their surroundings. All five sensory experiencesof living creatures are created through the reception of EM waves. The discovery of EM waves as the communication principle of the senses was made in the mid-1800s, and led for instance to the human voice being transformed into EM energy waves and broadcasted through wireless telephones or radio.

6.2. In 2014, scientific evidence proves sound is,in fact, EM energy waves

6.2.1. The fact that EM waves convey sound was discovered in the 19th century. This has meant a revolution for earthly communication through radio and telephony. Vocal sounds, i.e. EMwaves emitted from the vocal cords, could now be transmitted from aradio station to any place around the globe, and inside the receiver's auditory organ, the waves would then be transformed back into the original vocal sound. This revolutionary principle has been viewed as more of a technical discovery than a discovery of the peculiar natural principle that it is based on: EM waves. The theory thatEM waves of a certain wavelength being real light was quite simply accepted by the world of science as a fact. The questions of why EM waves exist naturally and where they come from had not yet beenanswered and remain unsolved to this day, as the physical sciences have been more interested in the particle theory than the EM wave theory. The most important clue to the understanding of EM waves is light. An understanding of light and vision will at the same time give a comprehensive, overall clue to all sensory experiences.

6.2.2. The principle and "technology" of sound was primarily developed by Maxwell (1831–79), Hertz (1857–94), Marconi (1874–1937) and Tesla (1856–1943). This led to EM waves being used in a practical sense for radio transmissions from 1896 onward. The explanation of EM waves as an energy principle was however replaced by a particle theory at the beginning of the 20th century;a theory mainly advocated by Einstein.

6.2.3. A particle theory was easier to handle, which led to the abandonment of the already established and convincing evidence of EM waves as the atomic language. Since then, science has tried to solve the problem by considering both theories as valid, which may appear as a practical solution, but as a matter of factis untrue because it contradicts facts, and instead creates confusion in the scientific subject matter. Particle theory of nuclear energy communication is certainly beneficial as an illustrative explanation and for mathematical quantification, but is not a correct description of reality – which is atoms that emit EM energy waves.

6.2.4. In 2014, a scientific discovery at Chalmers University of Technology in Gothenburg, Sweden, proved that soundis EM waves. The assumption and the evidence that the principleof light is indeed EM waves, which are perceived as light when they reach the eye, are based on historical observations, facts and circumstances particularly linked to light. The universal nature of the EM wave principle helps to turn EM waves as the mediating medium of "sense data" into a convincing explanation for all sensory experiences. A press release sent out by Chalmers University of Technology in Gothenburg in 2014 also describes a scientific study of the "principle of sound", which prompts a theory of sound, in fact, being EM waves. Because the research done at Chalmers has resulted in a discovery of sense data transmission through EM waves,which largely matches the general theory of all sensory experiences,it is reasonable to view this discovery as a confirmation of the EM wave theory, and that EM waves are the mediators of all sensory transmissions and impressions.

6.2.5. "The breakthrough: They talk to atoms".

In 2014, Chalmers University of Technology announced the discovery of atomic communication through sound:

"It is possible to use sound to communicate with atoms. This has been ascertained by scientists at Chalmers, the first ever in the world to accomplish this. The result can lead to extremely fast computers. That atoms and light interact with each other is a well-known fact, this research subject is called quantum optics. Per Delsing, Professor of Experimental Physics and his colleagues at Chalmers are one of the most prominent research teams in the world when it comes to quantum optics. They are now reporting a uniqueaccomplishment as they have managed to communicate through sounds with a man-made atom.

'We have opened a new door to the world of quantum physics,where we can listen to and talk to atoms. Our long-term aim is to tame the quantum physics so that we can utilise its laws, for instance in super-fast computers', says Per Delsing."*(Ny Teknik magazine)*

6.2.6. Atomic communication through EM waves is proven. What humans perceive as sound is always EM waves of different frequencies, depending on the nature and strength of the sound.But sound is not the only thing that is, in fact, EM waves;the same applies to light, taste, smell and touch, i.e. all five sensory experiences. What the Chalmers scientists discovered, that it is possible to communicate with atoms, generally confirms that it isthe communication of atoms through EM waves that convey the energy used for the sensory experiences of living species, and which the senses also receive. EM waves are not only the principle of atomic communication in nature, but also of all the sensory experiences of humans and animals, as well as of all communicationwith and experience of the outside world. It is, therefore, a completely logical and anticipated discovery, because human sensory functions are based on the same communication principleas for everything else in nature, which basically states that there is always communication going on between atoms, and always occuring through the fundamental energy language of atoms, i.e. EM waves.

6.3. Light, sound, colour, scent, taste and touchare all conveyed by EM energy waves

6.3.1. All five sensory experiences are created by the sensesreceiving atomic EM waves. Just as light and vision are sensory experiences based on EM waves, the other sensory experiencesuch as sound, taste, touch, and smell are all based on the same principle. All sensory experiences are conveyed and created byatomic EM waves, either as direct physical contact or through the air. The experience of taste takes place in the gustatory system, for instance, when EM waves are emitted from the atoms of the foodas it is processed in the mouth, while the experience of a scent is created in the olfactory system by EM waves emitted into the air from odour-creating matter, and the sense of touch by EM wavesat the point of physical contact. Atomic EM energy waves have a much greater role than previously assumed. In fact, EM waveshave become almost completely ignored, because the particle theoryof physics was incorrectly used as the explanatory model of the atom.

6.3.2. Each species interprets and experiences EM waves differently due to the unique construction of their senses and brains

Light/eyesight: EM waves from the atoms of a light source – director reflected – create an experience or perception of light insidethe eyes and brain of a living creature. The EM waves of light are reflected by the atoms of matter.

Sound/hearing: EM waves from the atoms of a sound source direct or reflected – create an experience or perception of sound inside the hearing organ of a living creature. For instance at physical impact, EM waves are emitted from a sound sourceand are conveyed via atoms of air to the hearing organ of a living creature. Audible sound that is conveyed as EM waves via atoms has, compared to the EM waves of light, a much lower energy level and can only be created within Earth's atmosphere. It is not possible to, for instance, communicate with normal voices on the moon,as the moon is surrounded by a vacuum and therefore cannot transmit the weak energy of the voice – contrary to the light energy of the sun and the other stars, which does not need any help from atoms to be transported through space.

Smell/scent: EM waves from the atoms of a scent source createan experience or perception of a scent inside the olfactory organ that a living creature can pick up on from a certain distance. Just like the EM waves of sound, the EM waves of scents cannotbe transported or perceived on, for instance, the moon, as it lacks atmosphere, i.e. atoms that make up the air. On Earth, the EM waves of scents are like the EM waves of sound: relatively energy-weak, with limited propagation. The propagation of both sound and scent is facilitated by the direction of the wind, because then the atoms of the air use their energy to convey EM waves and move with the wind.

Taste: EM waves emitted from what is ingested in the mouth is a direct transfer of EM waves from what has been ingested, to the tastereceptors of the mouth. The gustatory sense, just like the other senses, is connected to the brain where the experience is created.

Touch: EM waves are experienced when the superficial atoms of the skin of a living creature come into direct contact with other atom structures, or from a distance by the emitted EM waves of atoms such as cold or hot temperatures. The sense of weight occurs when planet Earth uses its greater energy to attract the body of a living creature.

7. LIGHT SOLVES THE MYSTERY OF GRAVITY

7.1. The current theory of gravity is based onan illusion

7.1.1. Gravity was previously called "the force of weight". This shows that the scientific understanding at the time was that gravitation, i.e. the force of attraction towards the surface of planet Earth, was assumed to be caused by real weight inside the matter. That each material or substance has a particular weightis only natural to man, because that is how man perceives every loose object. There has therefore, at least for this reason, been no point in questioning the principle of weight. From a scientific standpoint, however, the problem is that the weight of a substance does not actually exist the way man perceives it.

7.1.2. There is no such thing as real weight. All matter is actually weightless energy, and the "heavier" it feels, the more energy it contains. What mankind perceives as heaviness and weight, both for one's own body and other objects, is actually dueto the fact that all loose material units on Earth are electrically charged depending on the type of material and size, and that planet Earth is also charged. This causes a physical force of attraction, which means that smaller energy units are pulled towards Earth with a force equivalent to what man calls weight/ heaviness. This principle is totally logical and well-known within the world of science. All scientific progress needs a certain period of time to fall into place and fit into the bigger picture before the next logical correction can be made. What science has to do nowis to swap the theory of matter consisting of mass, for matter consisting of energy.

7.1.3. Still, to this day, the real nature of gravity is not clear. Newton (1642–1727) became the first person to study the phenomenon of everything falling to the ground from a mathematical perspective.He did not, however, question why this happens, but assumed that some real weight exists in all kinds of matter. Newton's lawof gravity thus became the foundation of natural sciences. Thereby, the mathematical principle of gravity was also scientifically determined, and the weight of a material or an object was accepted as a way to measure weight. This was an important, and at the time necessary, conclusion in an attempt to create alogic behind existence on Earth, but has meant a historical errorhas been accepted as a natural law because, at that time, it was not possible to draw the correct conclusion. What actually happens is that a greater energy unit attracts a unit of less energy.

7.2. Gravitation is caused by the energy of an object in relation to the energy of the planet

7.2.1. The mystery of gravitation[56]. Gravity has historically not been seen as a scientific problem, as the mass of matter has been assumed to constitute real mass with real weight. Scientists have known since 1785 that "Coulomb's law", a mathematical formula for the force of attraction between energy units, is mathematically identical to "Newton's law", which is an equation to calculate the gravity between so-called physical objects. Newton's law is not, however, evidence to support that a physical mass is a gravitational force, which Coulomb's law is for energy as a gravitational force.

**7.2.2. Gravity according to Coulomb´s law was

explained unwittingly in 1919 by Einstein and Eddington. This happened at a natural demonstration relating to a solar eclipse.The expectation was to prove that gravity is caused by the actual mass weight of an object in accordance with the atomic particle theory, i.e. that atoms contain mass. It was actually found that the expected gravity occurred, i.e. the light of a star diverted towards the sun, currently hidden behind the moon. This was then interpretedincorrectly as proof of both the star and the sun containing mass. Today, we know that since light is energy in the form of EM waves, the starlight cannot have deviated towards the sun due to gravity caused by the attraction between two masses. Gravity must, instead, be due to the attraction between two separate energies –for instance starlight and the sun, both consisting of energy. Theold theory of weight being due to two masses of matter causing attraction, according to the existing law of gravity, must therefore be corrected, and instead teach that gravitation is caused by the energy relation between two bodies, where a greater energy attractsa smaller.

7.3. Real weight does not exist

7.3.1. Nothing in the universe weighs anything. Weight, or gravitational force, is a force of attraction, a pull on an object or a living being that happens to be on the surface of that planet Weight can be described as a function of the difference between the energy content of a loose object compared to the energy content of the planet. The smaller the difference is, the stronger the attraction is to the larger object. Nature, planet Earth, and the entire universe are therefore not made up of matter consisting of atoms with a mass thathas its own weight, but of atoms with energy.

7.3.2. The atomic energy concept was presented in 1924by de Broglie, for which he was awarded the Nobel Prize for Physics in 1929. Nevertheless, the idea of matter having a mass of real weight has remained, because this is how the world is naturally perceived by man. It may seem strange thatde Broglie's qualified and accepted expertise on the subject failed to change the antiquated idea of the atomic constitution, from mass to energy, even though no mass could ever be found that was supposedto explain gravity. Instead, man has carried on with his search for theassumed atomic mass in the form of particles believed to exist insidethe atom. At a cost of 6.5 billion U.S. dollars, the "Large Hadron Collider" was built by CERN in Switzerland, with the aim of trying to determine whether mass really does exist in atoms. No mass was found. The reason is that real weight does not actually exist. Instead, the so-called weight of the energy content of bodies, i.e. that a bodyof greater energy attracts a body with less energy, is the explanation for gravity.

7.3.3. Until the 21st century, the world of science has considered atoms to contain particles with mass. The "Large Hadron Collider" has been representing the hope of being able to confirm that particles really do exist inside atoms. The troublesome aspect is that the whole project is based on the theory of mass existing, and therefore it is necessary to find particles to corroborate this. The "Standard Model" in particle physics predicts that fundamental particles should be found, of which the "Higgs particle" is the most famous and which should explain the construction of mass, and why matter has mass. So far, they have only found particles without mass. Everything suggests that the question of whether atoms contain particles in orbit or whether they consist of energy can be seen as advocating the latter. How atomic energy is created or added are left unanswered.

7.4. Everything that exists consists of atoms with energy

7.4.1. Atoms consist of different energy levels. Everything that can be physically defined in the human world consists of atoms containing different amounts of energy. What has surprised mankind is that the most energy-rich atoms contain such a vast amount of energy, which has been confirmed by the atomic bomb. From approximately 1900 onward, modern-day science first thought that atoms contained particles in orbit according to the model created to explain the energy capacity of atoms. When it later became apparent what an incredible amount of energy atoms actually contained, it was obvious this was a completely different interior structure than the scientists could ever have imagined.

7.4.2. How to explain atomic energy. With the insight and evidence of the vast energy content within atoms, a theory was created based on the assumption that an original, gigantic energy unit containing all stars and planets in the universe exploded at one point, and the parts spread out in the voidof the universe. These parts, in their turn, created all the celestial bodies that humans have discovered, but there areso many more, an infinite number, in an "everlasting" universe. The overall prerequisite for the existence of everything in the universe is energy, i.e. what created atoms in the first place. "The Big Bang" is the current theory of how all energy, i.e. matter, was once created, for some unknown reason, by an original explosion, and the energy has spread throughout the universe.

8. FINAL WORDS by the author

This reflection is not just about light. It is also about the people who participated fully or partially in this crucial scientific quest to find out what light really is. Thanks to the Internet with its amazing sourceof historical information, it has been possible to more or less conductinterviews with the ancient philosophers. It has been fascinating to get to know these wise, historical characters, and to piece together the jigsaw puzzle you might call the secret of light. Somewhat weirdto cast one's mind back to how this project and adventure began, and how I, without any prior knowledge in neither the subject of "light" nor the historical eras when the ancient philosophers lived, have had the great pleasure to get to know these important personalities to someextent. Hopefully, we have managed to combine historical knowledge with some understanding of the new world, in order to gain insights that might otherwise have remained hidden or forgotten.

I hope you have enjoyed it! I have.

Dag Landvik

9. REFERENCES

1) **The International Year of Light 2015**
Cited Jan 2018. Available from:
http://www.unesco.org/new/en/unesco/events/prizes-and-celebra-tions/celebrations/international-years/international-year-of-light/

2) **Klein, J. Martin. "Einstein and the development of quantum physics",** *Einstein: A Centenary Volume, The First Phase of the Bohr-Einstein Dialogue, p 138. Harvard University Press, 1979.*

3,4) **Calvino, Italo. "Mr. Palomar"**

Cited Aug 2018. Translated by William Weaver. 130 pp. New York: A Helen and Kurt Wolff Book/Harcourt Brace Jovanovich. 1985.

5) **Fermi National Accelerator Laboratory. Pages of Light**
Cited Jan 2018. Available from: http://home.fnal.gov/~pompos/light/light_page0.html

6) **How Light Works**
Cited Jan 2018. William Harris & Craig Freudenrich, Ph.D. "How Light Works" 10 July 2000. HowStuffWorks.com. <https://science.howstuffworks.com/light.htm>

7) **The Glitter Path: An everyday life phenomenon relating physics to other disciplines**
Cited Jan 2018. International Newsletter on Physics Education. October 2004.

8) **Encyclopedia Britannica, Light**
Cited Jan 2018. Available from: https://www.britannica.com/science/light

9) *What Is Electromagnetic Radiation?*
Cited Jan 2018. Available from: https://www.livescience.com/38169-electromagnetism.html

10) *What is an atom? Live Science.*
Cited Jan 2018. Available from: https://www.livescience.com/37206-atom-definition.html

11) *Helios the Sun-God*
Cited Jan 2018. Available from: http://www.talesbeyondbelief.com/myth-stories/helios-the-sun-god.htm

12) *Thales of Miletus: One Of The Famous "Seven Sages Of Greece" Who Predict-ed A Solar Eclipse*
Cited Jan 2018. Available from: http://www.ancientpages.com/2016/08/12/thales-of-miletus-one-of-the-famous-seven-sages-of-greece-who-predicted-a-solar-eclipse/

13) *Seven sages*
Cited Jan 2018. Available from: http://www.livius.org/articles/people/seven-sages/

14) *The story of mathematics*
Cited Jan 2018. Available from: http://www.storyofmathematics.com/greek_pythagoras.html

15) *Vanderwerf, D F. The Story of Light Science: From Early Theories to Today´sExtraordinary Applications.*
Austin TX: Springer, 2017.

16) *Empedocles Greek philosopher and scholar*
Cited Jan 2018. Available from: https://www.britannica.com/biography/Empedocles

17) *Walther Ziegler Plato in 60 Minutes*

Cited Jan 2018. Available from:
https://books.google.se/books?id=7_qnDAAAQBAJ&pg=PA2&lp-
g=PA2&dq=platon+english+lan-
guage&source=bl&ots=6s1YUTj4Hi&sig=g1VV_xT08YZg-

tYt-e7bYTh6snIE&hl=sv&sa=X&ved=0ahUKEwjDw66VmPbYAh-
VGXCwKHbVJB- s44ChDoAQgmMAA#v=onepage&q=platon%20eng-
lish%20language&f=false

18) *Aristoteles*
Cited Jan 2018. Available from: http://www.xtec.cat/~jllort1/biolegseuropa/ar-
istoteles_eng.htm

19) *Heath, T. Sir. Aristarchus of Samos The Ancient Copernicus.*
Cited Jan 2018. Available from: https://www.amazon.com/Aristarchus-Samos-
Ancient-Copernicus-As-tronomy/dp/0486438864#reader_0486438864

20) *Nicolaus Copernicus*
Cited 2018 Jan. 19. Available from: https://www.biography.com/people/nico-
laus-copernicus-9256984

21) *Galileo Biography*
Cited Jan 2018. Available from: https://www.biography.com/people/galileo-
9305220

22) *What is Huygens' principle?*
Cited Jan 2018. Available from: http://www.physlink.com/education/askex-
perts/ae471.cfm

23) *Can we explain Huygens' principle taking into ac-*
count Maxwell's predic-tions?
Cited Jan 2018. Available from: https://phys-
ics.stackexchange.com/questions/91627/can-we-ex-plain-huygens-
principle-taking-into-account-maxwells-predictions

24) **H. Joachim Schlichting "The Glitter Path: An everyday life phenomenonrelating physics to other disciplines"**
Cited June 2018. International Newsletter on Physics Education. October 2004

25) **Empedocles**
Cited Jan 2018. Available from: http://www.crystalinks.com/empedocles.html

26) **Leucippus and Democritus**
Cited Jan 2018. Available from: http://www.encyclopedia.com/humanities/encyclopedias-almanacs-tran-scripts-and-maps/leucippus-and-democritus

27) **Theories of light; theories of vision**
Cited Jan 2018. Available from: http://people.brandeis.edu/~sekuler/SensoryProcessesMaterial/Alha-zen.html

28) **Alexander the Great Learning from Aristoteles**
Cited Jan 2018. Available from: https://www.awesomestories.com/asset/view/LEARN-ING-FROM-ARIS-TOTLE-Alexander-the-Great

29) **On Colors Theophrastus**
Cited Jan 2018. Available from: http://trisagionseraph.tripod.com/Texts/Colors.html

30) **War Machines of Archimedes**
Cited 2018 Jan. 27. Available from: https://explorable.com/archimedes-war-machines

31) **Ancient Visions**
Cited 2018 Jan. 27. Available from: http://www.newworldencyclopedia.org/entry/Ptolemy

32) **Galen Facts**
Cited 2018 Jan. 27. Available from: https://www.iep.utm.edu/galen/

33) Al-Kindi. Stanford Encyclopedia of Philosophy
Cited 2018 Jan. 28. Available from: https://plato.stanford.edu/entries/al-kindi/

34) Alhazen: Early experiments on light
Cited 2018 Jan. 28. Available from: https://www.vision-learning.com/en/library/Hidden/59/Alha-zen:-Early-experiments-on-light/170

35) On Light or the Beginning of Forms. Robert Grosseteste
Cited 2018 Jan. 28. Available from: http://inters.org/grosseteste-on-light

36) Today in science: Johannes Kepler
Cited 2018 Jan. 28. Available from: http://earthsky.org/human-world/johannes-kepler-birth-day-dec-27-1571

37) New World Encyclopedia. Pierre Gassendi
Cited Jan 2018. Available from: http://www.newworldencyclopedia.org/entry/Pierre_Gassendi

38) Treatise on Light by Christiaan Huygens
Cited Jan 2018. Available from: https://ebooks.adelaide.edu.au/h/huygens/christiaan/treatise-on-light/chapter1.html

39) Maxwell discovers light is electromagnetic waves
Cited Jan 2018. Available from:
https://www.google.se/search?source=hp&ei=C81tWsDrGsLosQG-84ZmQAQ&q=maxwell+light+is+an+electromagnetic+wave&oq=Maxwell.+ligtht+is+&gs_l=psy

-
*ab.1.0.0i13i30k1.20667.43333.0.47361.23.20.1.2.2.0.108.1765.16j4.20.0.
...0...1c.1.64.*

psy-
*ab..0.22.1782...0j0i131k1j0i19k1j0i30i19k1j0i22i30k1j0i22i10i30k1j33i22i2
9i30k- 1j33i160k1.0.jLYo4IN9F50*

40) ***The Problems of Philosophy by Bertrand Russell***
Cited Jan 2018. Available from: https://wiki.eecs.yorku.ca/course_archive/2014-15/F/4412/_me- dia/the_problems_of_philosophy_by_bertrand_russell.pdf

41) ***Fredriksson, I. (ed) The Mysteries of Consciousness: Essays on Spacetime, Evolution and Well-Being. Jefferson, NC:*** *McFarland 2015. p 21-54*

42) ***The quantum mechanical model of the atom***
Cited Jan 2018. Available from: https://www.khanacademy.org/science/physics/quantum-physics/quantum-numbers-and-orbitals/a/the-quantum-mechanical-model-of-the-atom

43) ***The Wave Theory of Light: Definition & Evidence***
Cited Jan 2018. Available from: https://study.com/academy/lesson/the-wave-theory-of-light-defini-tion-evidence.html

44) ***Newton's Particle Theory of Light***
Cited 2018 Jan. 19. Available from: http://galileo.phys.virginia.edu/classes/609.ral5q.fall04/Lec-turePDF/L20-LIGHTII.pdf

45) ***What Is a Photon? Photons Are a "Bundle of Energy"***
Cited Jan 2018. Available from: https://www.thoughtco.com/what-is-a-photon-definition-and-proper-ties-2699039

46) ***Light as electromagnetic radiation***
Cited Jan 2018. Available from: https://www.britannica.com/science/light/Light-as-electromagnetic-ra-diation

47) ***Newton's Particle Theory of Light***
Cited 2018 Jan. 19. Available from: http://galileo.phys.virginia.edu/classes/609.ral5q.fall04/Lec-turePDF/L20-LIGHTII.pdf

48) ***Rosenblum, B and Kuttner, F. Quantum Enigma.***

Cited 2018 Jan. 19. Available from: Second Edition, Oxford: Oxford University Press: 2011. p 75 – 85.

49) **Wavelength**
Cited Jan 2018. Available from: http://searchnetworking.techtarget.com/definition/wavelength

50) **Nikola Tesla: The Guy Who DIDN'T "Invent Radio"**
Cited Jan 2018. Available from: http://earlyradiohistory.us/tesla.htm

51) **Is there an equivalent of the sonic boom for light?**
Cited Jan 2018. Available from: http://math.ucr.edu/home/baez/physics/Relativity/SpeedOfLight/ cherenkov.html

52) **The Thought Stash Science, skepticism, astronomy and whatever else wefancy writing about**
Cited Jan 2018. Available from: https://thethoughtstash.wordpress.com/2011/01/03/how-eddington-demonstrated-that-einstein-was-right/

53) **The Large Hadron Collider**
Cited Jan 2018. Available from: https://home.cern/topics/large-hadron-collider

54) **Columb's law**
Cited Jan 2018. Available from: http://www.physicsclassroom.com/class/estatics/Lesson-3/Coulomb-s-Law

55) **The quantum mechanical model of the atom**
Cited Jan 2018. Available from: https://www.khanacademy.org/science/physics/quantum-physics/quantum-numbers-and-orbitals/a/the-quantum-mechanical-model-of-the-atom

56) **The history of gravity, by Hobie Thompson and Sarah Havern**
Cited Jan 2018. Available from: https://web.stanford.edu/~buzzt/gravity.html

BIBLIOGRAPHY

Amoroso, R.L. (ed.) Complementarity of Mind and Body: Realizing the Dream of Descartes, Einsteinand Eccles, New York: Nova Science, 2010.

Chalmers, D. The Conscious Mind Oxford: University Press, 1996.

Fredriksson, I. (ed) Aspects of Consciousness: Essays on Physics, Death and the Mind. Jefferson, NC:McFarland 2012.

Fredriksson, I. (ed) The Mysteries of Consciousness: Essays on Spacetime, Evolution andWell-Being. Jefferson, NC: McFarland, 2015.

Messiah, A. (1961), Quantum Mechanics, Vol. I and II Amsterdam: North-Holland PublishingCompany, 1961.

Pais, A. 'Subtle is the Lord ...' The science and the life of Albert Einstein Oxford: Oxford UniversityPress, 1982.

Penrose, R. (2005), The Road to Reality (Vintage Books, London).

Rosenblum, B and Kuttner, F. Quantum Enigma. Second Edition, Oxford: Oxford University Press, 2011.Russell. B. A History of Western Philosophy. New York: Simon and Schuster, 1945.

Stapp, H.P. Mind, Matter and Quantum Mechanics Berlin: Springer-Verlag, 1993. Wigner, E. Symmetries and Reflections Bloomington: Indiana University Press, 1967.

Wilber, K. Quantum Questions: Mystical Writings of the Word's Great Physics. Contains contributionsfrom such leading physics as Heisenberg, Schrödinger, Einstein, de Broglie, Jeans, Planck, Pauli and Eddington. Boulder, CO: Shambhala Publications, 1984.

www.ingramcontent.com/pod-product-compliance
Lightning Source LLC
Chambersburg PA
CBHW022053050726
47591CB00002B/521